SMALL-SCALE MINING
A guide to appropriate equipment

**Prepared by James F. McDivitt, Dennis Lock
and others, on behalf of the Association of Geoscientists
for International Development**

Intermediate Technology Publications in association with
the International Development Research Centre
with assistance from the Commonwealth Science Council

Intermediate Technology Publications Ltd,
103-105 Southampton Row, London WC1B 4HH, UK

Reprinted 1993

ISBN 1 85339 012 7

CSC(90)WHR-12
TP 262

Printed by the Russell Press, Radford Mill, Norton Street,
Nottingham NG7 3HN. Tel: (0602) 784505,

Dedication

In the Autumn of 1978 Tom Wels contacted ITDG and enquired modestly whether there was any way in which an engineer, recently retired from the mining industry, could help the organization which he had long admired.

This was the start of a relationship to which Tom brought an enormous amount of time, energy and enthusiasm, and out of which grew ITDG's Small-Scale Mining programme. He was no mean engineer either — he had ended his business career as a Fellow of the IMechE and Deputy Chairman of Seltrust Engineering. In his kind and patient way, he provided guidance and wisdom on many aspects of the Group's work, and made many friends in the process.

Sadly, he was struck down by illness and subsequently died in September 1988. His family asked that memorial donations be made to ITDG, and that generous gesture has helped towards the publication of this book. *Small-Scale Mining: a guide to appropriate equipment* was one of Tom's earliest projects with ITDG, and it is fitting that it is dedicated to his memory.

Contents

General introduction

The Association of Geoscientists for International Development (AGID) received a grant from the International Development Research Centre of Canada to prepare a guide to equipment suitable for use in small mines. This project, which has been carried out in co-operation with the Intermediate Technology Development Group (ITDG, founded by E.F. Schumacher) aims to bring together basic information on equipment for all aspects of mineral development. The subject coverage includes exploration, surveying, sampling, analysis and testing, drilling, alluvial mining, underground mining, sorting, crushing, grinding, processing, materials handling and transportation, pumping, ventilation, power supply, general purpose equipment and safety.

Small-scale mining is difficult to define precisely, but for the purposes of the equipment in this guide we have taken an approximate upper limit of 100,000 tonnes of ore produced per year. This covers a wide range, extending from the very small, mostly hand-operated undertakings, upwards to mines which are semi-mechanized and (particularly in developed countries) some mines which may be fully mechanized.

Again for our purposes, small mines would normally be those with limited proven ore reserves. In many cases such mines would not have the technical staff to prove additional reserves. They would have limited access to capital and could be expected to operate with a relatively short life expectancy. The technology employed tends to be simple (pitting and trenching rather than diamond drilling, for example). The equipment must be relatively inexpensive and easy to operate and maintain.

As one approaches the upper limits of the group of mines classed as small scale, some of the characteristics of medium scale or large mines become evident. There is better access to capital and technical staff, and there is likely to be a high degree of mechanization (usually with smaller versions of the equipment used in really large mines).

Thus the change from very small or artisanal through small to medium and large is very gradual. At some place along this gradient there is an ill-defined point at which a mine's characteristics or philosophy move from small scale to medium scale. This will vary from country to country, from mineral to mineral and even, perhaps, from mine to mine. In all cases this transition point can be related to equipment and the way in which it is used. The objective of this Guide is to identify and provide basic information on equipment that will help small-scale miners to move up the gradient, thereby improving the effectiveness of their operations while helping to relieve the harshness of an industry which, in many cases, still relies heavily on physical labour.

In identifying equipment which is suitable for this range of users we can define a number of broad categories. These overlap and may contain duplications. The categories are:

1. *Standard equipment*, which can be used in either small or large mines. Much mining equipment falls into this category, including trucks, dozers, scrapers and other standard earthmoving equipment, drills, surveying instruments, some concentrating and pumping equipment, and so on. Units come in many sizes, but there are standard models which can be used in small mines producing around (say) one hundred to three hundred tonnes per day.

2. *Equipment specially manufactured for small mines*. In many cases this would be equipment which owes its design to an earlier time, such as ore cars designed to be pushed by hand or pulled by animals and semi-mechanized tools and equipment which combine manual and mechanical power.

There is special interest within this category in equipment from China and India, since these are among the few countries in the world which have the combination of a mining industry creating a large enough demand for such equipment and a manufacturing industry capable of supplying the demand. Brazil is another example.

In China, for example, some 5m people work in 200,000 small-scale mines (of which 80,000 are coal mines). A whole industry exists to provide equipment for this sector and, since such equipment may be particularly suitable for small mines in developing countries, China is anxious to widen its market. There are a number of Chinese items in this Guide, but this number could usefully be expanded in later editions.

3. *Appropriate or intermediate technology equipment*, which often can be made locally at relatively low cost, based on existing designs. Here also many designs originate from the past, going back to the last century or even earlier when all mines were small mines. The category includes slushers, drill wagons, various types of crushers, grinders and mills. This equipment can be made much more effective with modern supporting materials, components and drive units. For example, small generators and electric motors, lightweight plastic pipes and cables can greatly increase the effectiveness of yesterday's technology, making the equipment very useful in today's small mines.

Also within this category, but on a slightly different level, there are no doubt many examples of appropriate technology which were developed many years ago in one part of the world, and have since been superseded by standard mechanized pieces of equipment without ever having spread to other parts of the world. For example, there is a type of eccentric animal-powered grinding wheel which was once used extensively in Chile that is not seen in other areas. Another example comes from early placer mining days in Canada's Yukon, where a technique called the 'wing dam' was developed, that allowed miners to move their operations along the river. This technique also may not be known in other parts of the world. A very simple example is the use of bath towels to line sluiceboxes (little more than a modern adaptation of the ancient 'Golden Fleece') which is widely used in some parts of the world. In countries such as China or Brazil there must be many examples of mining equipment or techniques which have been developed locally that could be of use to miners in Africa or Latin America.

4. *Equipment of high capital cost*. There is, of course, much equipment which would be of value to small miners but is out of their range because of its scale, or out of their reach because of its price. This might be processing plant (such as mills or separation units) or it could be equipment for quality control, testing, and so on. In such areas equipment can often be shared or used on a co-operative basis, with government support in many instances.

Information has been gathered on equipment from all these categories, from a wide variety of sources. Perhaps the most useful and interesting items are those concerning old techniques, using devices which are simple to construct and maintain, but which (for one reason or another) have never been adapted to take advantage of modern materials, or have never been exploited outside the limited regions in which they originated.

It is evident that small-scale mining has much in common with industries such as construction and agriculture. Many of the processes are rather simple and direct, involving earthmoving, breaking and sorting of materials, drainage, water and power supply, and the transportation of bulk materials over short distances. Thus small mining equipment must meet many of the same criteria as other intermediate technology equipment. These criteria, as adapted from *Appropriate Technology Sourcebook* (Volunteers in Asia, 1986) specify that the equipment:

— shall require small amounts of capital
— shall emphasize locally available materials
— shall be relatively labour intensive, but more productive than many traditional technologies
— shall be small enough in scale to be affordable to local groups
— can be understood, operated and maintained by local people without a high level of training
— can often be produced in local workshops
— shall be flexible and adaptable to local circumstances
— shall be in harmony with local needs, traditions and environment
— shall extend human labour and skills rather than replace human labour and eliminate human skills
— shall place emphasis on self-reliance and local production to meet local needs
— shall minimize the impact of infrastructure limitations and the shortage of highly trained manpower.

These criteria, emphasizing as they do simplicity and self-reliance, are difficult to build into the programmes of large organizations but are well suited to projects carried out by small groups. This is one reason why many activities related to the promotion of small-scale mining are being carried out by non-governmental organizations (NGOs) such as AGID and ITDG, although, based to some extent on the initiative of these NGOs, increasing interest is being shown by larger organizations. This was evidenced by the recent seminar on small-scale mining organized in Turkey by the United Nations, and is also demonstrated by projects related to small-scale mining under the auspices of the World Bank and some of the bilateral assistance programmes.

In compiling this first edition of the Guide, a large number of information sources have been contacted. These include equipment suppliers, government agencies, mining associations, appropriate technology groups, and assistance agencies working in related subject areas. Interest in the project has been good, with adequate response to our requests for information (although less overwhelming in breadth). Our initial target was to identify at least 100 items for inclusion in the Guide, and in the event there are approximately 150. While we cannot claim that our coverage is complete, and many items have yet to be identified (particularly on the local, regional and appropriate technology levels) it has been possible to identify the general range of equipment that is currently available, and thus to establish the basic pattern and outline for this and future editions of the Guide.

So far as gaps in equipment availability are concerned, it appears that much of the rather simple equipment which was used one hundred years ago (when almost all mines would have been classed as small by our standards) has been replaced by larger and more sophisticated units. At the same time, upgraded versions of this early equipment can be of much use in the small-scale mining sector. This might involve using small motors and plastics (as already mentioned) and even computers, which can give new life to old designs. Old equipment catalogues, advertisements in the journals, and information in textbooks from the turn of the century are all good places for starting the search for ideas.

There are some areas in which useful research could be done, and from which practical information could be acquired to build up the coverage for future editions of the Guide. Since this must necessarily take considerable time, organization and support, it was decided to publish this edition as early as possible, recognizing that it will serve as the basis for further work in the field. We hope that it will encourage readers to identify and bring to our attention other items or references which they feel should be included. On this level, there is special interest in equipment and techniques that are specific to a region, and in equipment which, although no longer in use, might be adapted and reintroduced. We urge readers to send information on such items to the editors for inclusion in a planned supplement to this volume. We are also interested in information on groups which manufacture or market the equipment included in this guide for inclusion in a revised list of sources of equipment. Please send information to ITDG at the address on page ii, marked for the attention of the Mining Programme Manager.

Small-scale mining, as viewed today, has many similarities to mining at an earlier time, but there are also many differences. Not the least of these are the changes which have taken place in approach and philosophy. A new generation is involved in the promotion of small-scale mining, looking upon this section as an area of growing importance in which significant improvements and developments are possible that will have a positive impact on world mineral supply, but also, perhaps more important, on the miners themselves and on the environment in which they work.

As evidence of the growing interest in small-scale mining, the International Agency for Small-Scale Mining was officially established on 31 March 1989. It is managed by an international board of 17 members, and has its headquarters in Montreal, Canada. The Agency, known generally as Small Mining International (SMI), is placing its initial focus on information services, co-ordination and linkage among national and international organizations working with small mines and mineral industries. In the long term it hopes to be able to provide other support services including research and training. It publishes a Newsletter available through the SMI office. Contact: Small Mining International (SMI), PO Box 6079, Station 'A', Montreal, H3C 3A7, Canada.

James McDivitt

Small-scale mining: a general review

The history of mining until this century was one of small-scale operations, often crude in terms of technology and hazardous to health and safety but nevertheless providing the necessary mineral raw materials for society. In contrast, resource development in the twentieth century has been marked by the growth of large mining utilizing economies of scale. In the past thirty years and through two United Nations Development Decades, the prevailing theories have emphasized the need for rapid industrialization, backed by the belief that the benefits from large operations would 'trickle-down' through a national economy. Accordingly, technical assistance programmes in mineral exploration and development, in institution building and training focused on the need for large-scale production, and this approach still dominates.

In recent years this philosophy has come under increasing criticism in developing countries. The gap between the richer and poorer nations seems to have widened; transnational corporations based in the Northern Hemisphere have grown to awesome levels of power and the industrialization that has taken place has too rarely benefited the majority of the people in the society. Many planners and theoreticians have thus begun to emphasize a 'basic needs' approach to raising the quality of life and increasingly the focus has been on the growth of a self-reliant economy 'de-linked' from the North, on rural development and on the utilization of appropriate technologies.

It is natural, therefore, that the interest in minerals has begun to return to the idea of small-scale operations. The main lead came from the United Nations in a major survey of small mining in developing countries (Skelding, 1972). This was followed in 1978 by a United Nations and Mexican Government international conference in Jurica, Mexico, on *The Future of Small-Scale Mining* (Meyer and Carman, 1980). Even in an industrialized country like Canada where mining is a major industry, there have been serious calls for re-emergence of the small mining enterprise and the junior mining company (Freyman, 1978; Kalymon et al., 1978; Mineral Resources Group, 1978; Wojciechowski, 1979).

Another important event has been the move into this field of the UK-based Intermediate Technology Development Group set up in the 1960s by E.F. Schumacher, the world-famous author of *Small is Beautiful*, in order to further the application of appropriate techniques. The ITDG mining programme operates an industrial enquiry service, sponsors field visits and assists small businesses in acquiring and adapting processes and equipment, even to the pilot commercial stage.

What is small-scale mining?

There are many definitions of small-scale mining operations (Argall, 1978). Some utilize the number of people employed, others the size of the concession area, the size of reserves, the productive capacity, productivity of labour, the gross annual income, the degree of capitalization or mechanization, the continuity of operations or even the requirements of mine safety legislation plus various combinations among the foregoing. Of course, as the UN study pointed out, what would be considered a large mine in one country (e.g., Kilembe in Uganda) might well be viewed as small in another (Canada). Moreover, very different considerations must apply to high bulk/low value deposits such as bauxite as compared to low bulk/high value ones such as gemstones. In an extreme case, in a personal communication

Macdonald has said that with most rutile-bearing beachsands of Australia which characteristically grade less than one per cent TiO_2, a daily throughput of about 10,000 yards is close to the lower economic limit. Marinovic in Meyer (1980) provided data for 1976 showing that the 1,065 small copper producers shipping ore to the concentrators of the Mexican government agency Empresa Nacional de Mineria had an average daily output of 5 tonnes of run-of-mine. This is based on a working year of 300 days, which is highly theoretical as evidently many mines operated only sporadically.

In effect, nations will develop their own definitions as to what constitutes smallness depending upon their mixes of sociological, geographical, financial and technical factors. But from a global viewpoint, the only feasible measuring rod is tonnage. The Jurica Conference in 1978 was unable to reach a consensus but as an offshoot of that meeting a considerable body of work has been published which has been drawn together by Noetstaller (1987, Table 1).

Table 1. Classification values for small-scale mining, expressed in tonnes per year of run-of-mine ore

below 50,000	United Nations	1972
below 50,000	P.C. Kotschwar	1986
below 100,000	J.S. Carman	1985
below 100,000	G.F. Leaming	1983
below 100,000	D.N. De Bord and W.G. Mikutowicz	1981
below 100,000	U.S.B.M.	1983
from 20,000 to 200,000	D. Ingler	1983
below 150,000	*Mining Magazine*	1986
below 50,000	J.C. Fernandez	1983
below 60,000	G. del Castillo	1980

Ingler's figures are not relevant in this instance as they are probably designed to range beyond artisanal mining through small scale in the Third World to what is judged to be small scale in the industrial nations. The other estimates average out to a consensus of less than 100,000 annual tonnes as a logical global figure. It encompasses the small mines of Australia, South Africa, the USA and such which may produce up to 200,000 tonnes per annum of run-of-mine and the producers of the developing lands where, with all but low unit value bulk minerals such as sand and gravel, a mine extracting 25-30,000 tonnes yearly would be rated as medium sized.

Table 2, overleaf, was prepared to fill a gap. For many years it has been widely voiced that the small miners of the world account for 10 per cent of world output of non-fuel minerals. The figure was too neat to be believable. Accordingly the detailed study establishes that at least 16 per cent of the value is contributed by small miners. The words 'at least' are used advisedly because almost one-third of the minerals involved having their share rated as negligible fails to take into account the great number of small entrepreneurs throughout the world who seize upon opportunities to mine almost any mineral. Much of such production is on a toll basis and thus tends to enter into national statistics to the credit of larger operators.

General characteristics of small-scale mining

The typical mine operating on a small scale in the developing world is a producer, often sporadically, of limited amounts of mineral from deposits with few known ore reserves and of a

Table 2. Estimated value of the small mining sector in the production of non-fuel minerals in 1982

Mineral	Gross Value of Output ($ millions)	Share of Small-scale Mining (%)	Gross Value of Small-scale Mining ($ millions)	Price (US$)	Quantity (Thousands)
Antimony	126	45	57	1.07/lb	59 st
Asbestos	1,444	10	144	335/mt	4,311 mt
Barite	300	60	180	38/st	7,887 st
Bauxite	3,008	Negligible		40.42/mt	74,441 mt
Beryllium	38	100	38	6.30/lb	3 st
Bismuth	15	Negligible		1.87/lb	4 st
Boron	778	Negligible		311/st	2,503 st
Bromine	620	Negligible		0.75/lb	413 st
Cadmium	39	Negligible		1.11/lb	16 mt
Chromite	633	50	316	58/st	10,907 st
Clays	2,592	75	1,944	varies	149,803 st
Cobalt	675	10	68	12.50/st	27 st
Columbium	97	Negligible		3.04/lb	16 st
Copper	12,812	8	1,025	0.73/lb	7,963 st
Feldspar	124	80	99	33/st	3,745 st
Fluorspar	745	90	670	149/st	5,003 st
Gold	16,060	10	1,606	376/tr. oz	42,713 tr. oz
Graphite	221	90	199	364/st	607 st
Gypsum	682	70	477	8.46/st	80,616 st
Iron ore	32,638	12	3,917	41.72/lt	782,302 lt
Lead	1,977	11	217	0.26/lb	3,450 mt
Magnesium	731	Negligible		1.34/lb	273 st
Manganese	1,634	18	294	66/st	24,754 st
Mercury	77	90	69	377/fl	204 st
Molybdenum	158	Negligible		7.90/lb	100 st
Nickel	4,512	Negligible		3.20/lb	705 st
Phosphate rock	3,788	10	379	31/mt	122,202 mt
Platinum group	1,801	Negligible		280/tr. oz	6,431 tr. oz
Potash	3,830	Negligible		146/mt	26,230 mt
Pumice	114	90	103	9/st	12,702 st
Salt	2,703	20	541	14.53/st	186,000 st
Sand and gravel	10,103	30	3,031	3.23/st	3,128,000 st
Silver	2,962	10	296	7.95/tr. oz	372,528 tr. oz
Stone	14,957	30	4,487	3.78/st	3,957,000 st
Sulphur	5,471	Negligible		108/st	50,660 st
Talc and pyrophyllite	182	90	164	24/st	7,595 st
Tin	3,118	15	468	5.87/lb	241 mt
Titanium	413	Negligible		84/st	4,922 st
Tungsten	272	80	218	5.67 lb	24 st
Vermiculite	51	90	46	90/st	564 st
Zinc	5,064	11	557	0.38/lb	6,047 mt
Totals	137,565		21,610		

character not amenable to mass mining. Output is won without much use of mechanical energy and by exploitation of labour under leadership that commonly lacks technical, managerial and business skills. Small mining can lead to serious wastage of non-renewable resources by high-grading but paradoxically it also enriches nations and the world as a whole by often playing the role of scavenger. Per unit of output it is a prolific employer of labour and though the terms may be harsh, the alternative may be slow starvation.

Several major mineral groupings are especially amenable to small operations: the pegmatite minerals (e.g., rare earths, lithium, mica, tin, tungsten), precious metals, placer deposits (gold, platinum) and industrial minerals. Small scale also generally requires surface or near-surface deposition, very little waste or overburden, uncomplicated metallurgy and relatively easy access (Skelding, 1972; Meyer, 1980).

Small-scale mining and the steps that lead to production generally involve the application of low, intermediate or appropriate technology in terms of 'low cost per workplace', whether as old-fashioned prospecting methods or as artisanal mining. Many of the techniques of exploration and exploitation have been virtually universal, as in placer mining, though unique or modern methods have sometimes been developed as, for example, in the case of dimension stone (UN, 1976), precious coral (Grigg, 1979), smelting (Summer, 1969) and gem cutting (Gubelin, 1968). Contributors to the Mexican Conference pointed out the potential application to small mining operations of heap

leaching (Kappes, in Meyer, 1980), of mobile ore dressing plants (Stigzelius, ibid), and even of small retorts for the treatment of oil shales (Savage, ibid).

On a global basis, the vast majority of small mines are worked by one or two people (placers), by a family or small gang (non-metallics, base metals), by various forms of leasing or tributing where the miners share the output with the owner of the mineral rights, by joint ventures of all sorts and co-operatives. Distribution and sales commonly involve public agencies which treat the raw ore, government buyers of gold and gemstones, private buyers of the same, often acting clandestinely, and metal merchants where the products enter world markets. Industrial minerals are on the other hand often processed and consumed locally.

As West and Colli (in Meyer, 1980) have pointed out, large-scale operations are less likely than small in the case of low net value minerals which cannot be transported far because of cost. Likewise, small high-grade deposits and those subject to large erratic shifts in market demand and price cannot generally sustain the high overhead costs of large operations, whereas the small operator usually has the flexibility which permits suspension of work when the times are bad.

Advantages of small mining operations

As a labour-intensive activity, small mining operations employ large numbers of workers, generally in areas remote from cities, especially where placer deposits are involved. According to

Argall (1978), 97 per cent of all operating mines in India are classed as small scale and employ nearly 50 per cent of the total mining workforce. Currently, there are about half a million small-mine workers in India. Elsewhere, small tin and tungsten operations in Rwanda have employed up to 11,000 workers; 30,000-40,000 people were involved in the early days of alluvial diamond production in Sierra Leone; up to 40,000 people worked in the small mines of Bolivia; 15,000-20,000 in the Central African Republic; 15,000 in Venezuela; and so on (Meyer, 1980). A particular advantage is that such artisanal mining can be pursued on a seasonal basis, geared for example to the manpower demands of agriculture.

Small operations often lead to the recognition of major deposits and many a large mine had its origins in small workings. It is worth noting in this respect that during the extensive mineral exploration in Canada in the 1950s smaller mining enterprises were highly effective, being responsible for 62 per cent of all economic finds, with the expenditure of less than 30 per cent of the total funds spent on metal exploration.

Small operations can form the basis for local processing and manufacturing industries, either on a small scale or as feeders to larger centralized plants. Examples include the artisanal cutting and polishing of gemstones, splitting of mica, the use of clay for bricks and ceramics, silica sand for glass, gypsum for cement and so forth. In such cases, the value added through processing is not 'exported' as in the case of many larger mining operations. Involvement in small-scale mining can also provide a practical way to familiarize people in remote areas with the workings of a money economy. And it can help to counter the deadly migration to already overcrowded cities.

The development of small-scale deposits is almost always accomplished much more rapidly than with large deposits and at a fraction of the cost. And despite popular mythology, the cost per productive unit will not be substantially greater than with a major operation. The law of diminishing returns tends to set in at an earlier stage than realized. A study made ten years ago on the cost of construction of zinc concentrators in Andean countries covered a range from 20-300 tonnes per day. Costs per daily tonne treated were halved at the 150 tonne level and thereafter the curve was virtually flat. Moreover, small mining generally avoids many of the big problems that often plague large-scale mining in developing countries — problems in financing, in dealing with the transnationals, in building extensive infrastructure, in acquiring and applying imported technology and in supplying the necessary expertise. In the case of governments bedevilled by such difficulties and unconvinced of the need for mineral development, the effect of developing a successful smaller deposit could be salutary. Likewise, the experience gained by young professionals and technicians in building up or operating a small mine could be vital. The environment is far more challenging than the bureaucratic atmosphere of the large, well-established operation.

Disadvantages of small operations

There are, as everyone involved with small operations knows only too well, serious drawbacks to small-scale mining. Small mining is a brutalizing business often characterized by vicious exploitation of labour. Less-than-subsistence wages, appalling working conditions, and a total disregard for health and safety are far too common, frequently resulting in a life expectancy of only 30-35 years for underground workers. Another example was provided quite a few years ago by the bringing into production of a large iron ore mine in India, the design of which embraced the last word in modern technology. Questions were raised regarding comparative costs vis-à-vis a similar deposit nearby which was being worked without mechanization. There was some consternation when it was shown that the costs of tyre wear and vehicle maintenance per tonne would exceed all costs per tonne extracted at the other operation for the simple reason that its wage rates were 20 cents per shift for men and 15 cents for women, although the women did the hardest work.

Small mining frequently leads to fractionation of ore bodies where a single deposit of several million tonnes, for example, has been so cut up in claim staking that a score of owners may work parts thereof at different times, rates, elevations, the whole totally uncoordinated and extremely wasteful. If and when such properties are taken over by a single agency, existing underground workings, instead of being a definite asset, constitute a major barrier to rational exploitation. Particularly with placers, the dumping of waste material on adjacent reserves may reduce them to a submarginal category (Brower, 1979).

The problems of controlling and regulating small mining industries can be immense. Where gems and precious metals are concerned, illicit operations are often widespread, smuggling is common and lawlessness prevails. An influx of buyers and middlemen of foreign origin can create awkward situations and even raise foreign policy issues of major concern to the host country. The extreme difficulty of monitoring small-scale production of virtually any mineral commodity means that revenues to governments are much less than they should be in terms of licence fees and taxes. Moreover, there are obvious problems in providing the necessary finance, in encouraging efficient processing and forward linkages through fabricating and manufacturing and in obtaining satisfactory marketing arrangements (see Chender, in Meyer, 1980).

In all of this there are vicious cycles at work where the inability of Government to control small mining or the reluctance of the local private sector to invest in small operations may result from the lack (or perceived lack) of substantial benefits to the investor and the nation. Funds to support field inspectors, to encourage research, to improve local expertise or to promote private investment are consequently lacking. It becomes effectively impossible to monitor and regulate small mining, and to demonstrate the benefits that could be derived. If it is worthwhile to break this cycle then the challenge is to devise ways — and strategies — of doing so.

J.S. Carman and A.R. Berger

References

Argall, G.O., Jr. (1978). Conference on the future of small-scale mining. *Important for the future*. UNITAR, New York.

Brower, J.C. (1979). Small-scale mining and economic aid in Bolivia. In *Natural Resources Forum*, United Nations, New York.

Carman, J.S. (1979). *Obstacles to mineral development: a pragmatic view*. Pergamon Press, Elmsford, New York.

Freyman, A.J. (1978). *The role of the smaller enterprises in the Canadian mineral industry with a focus on Ontario*. Ontario Ministry of Natural Resources.

Grigg, R.W. (1979). 'Precious corals: Hawaii's deep sea jewels'. In *National Geographic*.

Gubelin, E. (1968). *Die edelsteine der Indel Ceylan*. Kommision Ed. Scriptar, Lausanne.

Kalymon, B.A., P.J. Halpern, J.D. Quirin and W.R. Waters (1978). *Financing of the junior mining company in Ontario*. Ontario Ministry of Natural Resources.

Macdonald, E. (1983). *Alluvial mining: the geology, technology and economics of placers*, 3rd ed., Chapman and Hall, London, New York.

Meyer, R.F. and J.S. Carman (1980). *The future of small-scale mining*. McGraw-Hill, New York.

Mineral Resources Group (1978). *The decline of small mineral enterprises in Ontario*. Ontario Ministry of Natural Resources.

Noetstaller, Richard (1987). *Small-scale mining: a review of the issues*. World Bank Technical Paper No.75. Industry and Finance Series. The World Bank, Washington.

Skelding, Frank (1972). *Small-scale mining in the developing countries*. United Nations, New York.

Summers, R. (1969). *Ancient mining in Rhodesia and adjacent areas*. Memoir 3, National Museum of Rhodesia, Salisbury.

United Nations (1976). *The development potential of dimension stone*. Department of Economic and Social Affairs, New York.

Wojciechowski, M.J. (1979). *Junior mining in Canada: the problem of investment and securing fair returns*. Proceedings 6, Centre for Resource Studies, Queen's University, Kingston, Ontario.

Editor's introduction

This Guide has been compiled with the principal objective of giving practical information and suggestions to those seeking equipment for small-scale mining operations, especially in developing countries and in remote parts of the world.

Most of the entries list typical sources of commercially available tools, instruments and equipment needed for many aspects of prospecting, mining and mineral processing operations. We also give examples where locally available materials, including surplus and scrap items, can be used to make or improvise equipment that saves the cost or difficulty of buying commercially manufactured items.

How the entries are arranged

In general we have arranged entries in a sequence that approximates to the actual sequence of tasks experienced in mining. Thus we start with prospecting and surveying, which is followed by mine development and working before we list equipment for mineral processing. The remaining sections deal with equipment used more generally across all stages of mining and ore treatment and handling.

Some equipment does not fit neatly into one category. For example, the ubiquitous batea or pan would be regarded by many as a prospecting device, but it becomes a production tool for a small one- or two-man mining operation.

Complications arise when attachments are made for machines, so that a drill attachment might be bought for a dumper. The dumper will be found in Section 5, since we have classified it as a materials handling and haulage device. Its drill attachment is a surface development tool, and is therefore listed in Section 2.

Mine-made equipment

Almost all of the items listed in this Guide that can be produced at the mine or locally will be found in Section 4. They range from very primitive implements to slightly more complex designs needing the services of an engineering workshop.

These designs, ideas and practical tips have been collected from a variety of sources and they owe their origins to mines and miners from all parts of the world. Some of these items were first thought of many years ago, but they have stood the test of time and remain valid for application in remote areas where there is difficult access, no power supply and perhaps a lack of purchasing capital.

In order to describe and illustrate these mine-made devices we have sometimes drawn on reference material from past technical reports or textbooks. We have endeavoured to acknowledge all of these sources with the relevant entries.

No data have been given on the estimated costs of making any item. This is because circumstances differ so greatly from one place to another, and the costs must depend on what materials are available locally and on how much is paid for labour.

Equipment available for purchasing

All of the purchased equipment items are derived from information in brochures and catalogues supplied by companies wishing to be included, many of whom have expressed support for the aims of this Guide.

It has generally been possible to include only one or two representative items from each company's range. Many of the manufacturers and suppliers listed can offer a far wider range of goods than those shown here. If, therefore, you have a specific requirement that we have not covered, you will probably find that an enquiry to one of the contributing mining equipment suppliers will be worthwhile.

Even where the company approached cannot supply the idea needed they will often be a good starting point, since most firms who supply any equipment to mines have a good knowledge of trade sources and are willing to put buyers in touch with those who can supply. It should be said that we found firms very helpful in this respect when we were researching for the Guide.

Price information

Because much equipment pricing depends so greatly on the particular specification and optional accessories needed, and with the passage of time (and therefore cost inflation and international exchange rate movements) many suppliers have been reluctant to give price information on their products. Even where prices are mentioned in the Guide, these are given only for broad budget purposes.

Safety

Mining has not earned a reputation as being one of the safest occupations. This Guide offers information only on the sources and availability of equipment, with recommendations for its safe construction, installation or use. Neither the publishers nor anyone connected with the preparation of this Guide can be held responsible for any accident or loss that might arise in connection with any item of equipment listed.

Any mine operator who undertakes the local improvisation or manufacture of any item that we describe should ensure that it is properly designed and stressed, and made to be safe during its installation, operation and at all other times.

Note

Every attempt has been made to ensure accuracy of the details presented in this Guide, but doubtless changes will have occurred about which the compilers are unaware. We apologize to any reader to whom we may have given a false lead. A note will be made of up-to-date information which becomes available to ITDG.

It must be stressed that this Guide relies on information supplied by the manufacturers and that inclusion of an item is no guarantee of performance. Whilst every care has been taken to ensure the accuracy of the data in this Guide, the publishers and compilers cannot accept responsibility for any errors which may have occurred. In this connection it should be noted that specifications are subject to change without notice and should be confirmed when making enquiries and placing orders with suppliers.

Dennis Lock

Acknowledgements

The publishers would like to thank the Commonwealth Science Council and the Estate of Mr Tom Wels, as well as AGID and IDRC, for the generous support given to the preparation of this publication. Many individuals also helped to compile the book; it is not possible to mention them all but in particular thanks must go to Graham Kill, Ernest Hogg, Edmund Bugnosen, and Rosalind Patching.

1. Prospecting, assaying and surveying

Prospecting is the initial stage of any mining operation, regardless of scale. Its very purpose is to locate and identify potential mineral deposits. It is a high financial risk activity generally conducted with the least expectations and entails no immediate returns. The returns are only realized if and when a mineral discovery resulting from the activity has been put into viable mining operation. Results of prospecting work are a basis for subsequent mineral exploration and evaluation — yet another financially high-risk phase of the general mining process involving higher expenditure requirements. Considering this foremost role which prospecting takes in the development process of a mine, it must be approached with careful analysis, both financially and technically. Aside from wise budgetary control, proper choice and use of equipment must be exercised.

In the past, prospecting was a relatively small-scale and rather unsophisticated operation. Indeed some of the big mines of today are the results of expansions of small mines which evolved from the simple prospecting activities of the past. It is perhaps the only phase in the development process of a mine when even the crudest equipment developed by the industry is still found in wide use. The simple pan is an example of this. While modern equipment, such as the 'Gold Genie' (see Section 4) was developed for similar use, the pan is still used to complement it. High-tech precision surveying equipment has not yet succeeded in eliminating the Brunton compass from the belt of the prospector.

Prospecting activities are mostly carried out in remote areas. As such, the use of portable equipment is not only popular but often a necessity. This is demonstrated by the increasing use of mobile and portable drills. Most of the equipment is also designed to be lightweight for easy transport and, at the same time, rugged and hardwearing to withstand rough terrain and handling in the field.

Assaying is an important aspect of prospecting work. Samples collected in the field through test pitting, trenching or drilling need to be analysed and results relayed immediately to field personnel, as often the progress of prospecting work depends on the results of analysis. Results must also be very precise. This can be achieved by well-trained assayers and the use of proper and appropriate equipment. Small mines may normally have to rely on a mine assay laboratory even if this means transporting samples some distance, as putting up an on-site laboratory is often not viable. Alternatively, portable on-site analysers may be used, or sample preparation done in the field to reduce transport and handling costs.

Surveying requirements in a prospecting project may not require sophisticated equipment. A small-scale prospector would normally need a compass and tape-measure, altimeter and a good topographic map of appropriate scale. However, a bigger project may require actual survey work, such as the establishment of grid lines for geophysical and geochemical sampling and drill site stations. Under such circumstances, proper surveying equipment is necessary. A theodolite for this purpose is described in this Guide.

Small mines may still continue to be involved in prospecting activities worldwide. But as the need to prospect for deep-seated and unexposed deposits increases, the traditional prospecting methods, using simple equipment, may decline. In effect, the participation of the small mines will correspondingly decrease and prospecting activities will eventually be more the concern of government agencies and big mining companies. The position of the small mines therefore in the prospecting field will greatly depend on the development of appropriate, small-scale equipment needed in prospecting, as well as well-trained personnel.

Further developments of equipment for on-site analysis will gradually simplify assaying procedures and may eliminate costly sample transport, handling and preparations. Similarly, the surveying aspect of prospecting will be influenced by advances in regional geological mapping and map preparation worldwide, particularly in the developing countries.

Edmund Bugnosen

GOLD PAN/BATEA

A basic yet universal tool for prospecting for alluvial deposits, particularly those containing gold and other heavy minerals.

A batea can be made by cutting a disc from a circular metal sheet which is 1mm thick by about 450mm diameter. One method is to cut the disc from one end of an oil drum.

The dish shape is made by beating the disc with either a 2 lb (1 Kg) ball peen hammer or a 4 lb (2 Kg) coal hammer with slightly rounded faces. During this process, the disc can be supported on the square cut face of a suitable log, dishing being obtained by first placing a sandbag on the log.

The inevitable hammer dents can be taken out by finishing the beating over a dolly made by shaping the end of another log.

The inside surface should be slightly roughened.

Alternative method

Another simple method for improvising a batea, which is widely used in Africa, is to cut a disc of rubber, about 450mm diameter, from an old truck inner tube. This is then held in the palms of both hands.

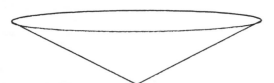

VENEZUELAN 'SOROCCO' SIEVES

These simple tools can be used for diamond/kimberlite prospecting in alluvial gravels.

For local manufacture, these sieves are circular, each about 60cm in diameter, with wood rims. In the top sieve the wire mesh needs to be of 6mm aperture, middle 4mm and bottom 1.5mm.

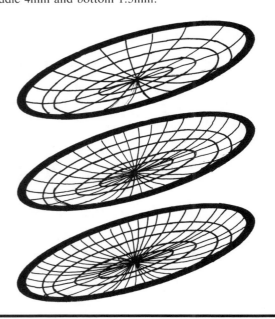

PORTABLE SPECTROMETERS

The manufacturer listed below supplies prospecting and surveying instruments. Listed here are just three useful examples of their wide range.

GAMMA RAY SPECTROMETER

This model, the GR-310, is a high resolution differential spectrometer suitable for geological mapping and reconnaissance surveys as well as site assay. It is a self-contained instrument with internal calibration reference isotope, (easy to calibrate during field use) and measurements of potassium, uranium and thorium are switch-selectable, as well as total gamma ray count. It has both a visual digital display and an audio tone.

GAMMA RAY SCINTILLOMETER

This model, the GR-101A, is a compact, lightweight, self-contained instrument. It is suitable for reconnaissance surveys, drill core analysis and general exploration.

Measurements are clearly displayed on a large 280 degree ratemeter, with an expandable full-scale range from 100-10,000 counts per second. An audio alarm provides automatic indication of gamma radiation above pre-selected background levels.

It is small and light enough to be carried from a belt, leaving hands free, or from a shoulder strap.

It is supplied with batteries, belt pouch, shoulder strap, operator's manual and test source.

PROTON PRECESSION MAGNETOMETER

The fully-portable model, MP-2, is a one gamma proton precession magnetometer for field survey use. It is very lightweight, has low power consumption and is durable and reliable in the field.

It is battery operated, has 1 gamma sensitivity and is accurate over a range of 20,000 to 100,000 gammas. It can operate in high gradients, to 5,000 gammas per metre. It reads out in 3.7 seconds, has a clear, LED digital display with complete test feature. It gives a digital readout of battery voltage and an indicator light warning of excessive gradient, ambient noise or electronic failure.

Source

E.G. & G. Geometrics   Telephone 416 661 1966
436 Limestone Crescent   Telex   0622694
Downsview
Ontario
M3J 2S4
Canada

QUARTZ SPRING GRAVITY METER

This gravity meter can be used in geodetic surveying and prospecting for, among other things, coal and metalliferous deposits.

Operating information

Type number	ZSM-3
Observing accuracy	plus or minus 0.03 mgal
Reading accuracy	plus or minus 0.1 div
Counter range	0.000 to 999.9 div
Counter constant	0.08 to 0.12 mgal
Reset range	more than 4000 mgal
Beam sensitivity	16 to 20 div
Drift	less than 0.3 mgal per hour
Weight	4.5 Kg

Accessories
Shipping container
Field carrying pack
Tripod
Tool case
This manufacturer also produces a thermostatically controlled unit with lower drift and larger counter range.

Source

Beijing Geological   Telephone 482261
  Instrument Factory   Cable   3654
2 Dong San Huan Bei Road
Beijing
People's Republic of China

GEOPHYSICAL PROSPECTING TOOL

A hand-held electronic prospecting tool for alluvial deposits, especially gold. The instrument is light and portable. It can be used for soils, gold bearing veins, river and sea beds, and for testing processing efficiency.

   The principle used is one of electro-chemical self potential with a statistical base, so that high concentrations of minerals are a result of many positive readings.

   Detection indication is by a display on the electronics box, and by audible signals received through headphones.

Operating information

Smallest detectable particles	300# (50 μm)
Low detectable grade	0.2 to 1 ppm
Probe length	1.2m, but 1.2m extensions can be added
Power	9 volt battery, type PP3
Battery life	50 to 80 hours

Source

Terra Probe   Telephone +461 825 7750
  International AB   Telex   12442
Box 24075
S-75024 Uppsala
Sweden

SLINGRAM ELECTROMAGNETIC SURVEYING EQUIPMENT

The slingram is an easily portable electromagnetic loop system comprising a transmitter and separate receiver. It can detect electrically conducting materials, such as ore bodies, water bearing fracture zones and aquifers. It has been used with good results in both mineral and water exploration in Europe, Africa and America.

Surveys are carried out by two people, each wearing one of the two units round their waist. A length of cable is stretched between the two. The parameters measured are the real component (in phase) and the imaginary (quadrature phase).

Good noise suppression rejects interference, even from power lines. Operation is simple, giving measurements that can be interpreted in the field.

Operating information

Frequency	800 Hz, 3.6 kHz, 18 kHz or 32 kHz
Coil separation	Vertical and horizontal coplanar or coaxial
Temperature	Operates from minus 30 to plus 60 degrees Celsius
Transmitter	
Battery	Rechargeable nickel-cadmium, allowing 10 hours continuous use per charge
Dimensions	575 × 575 × 31-90mm
Weight	7Kg

Receiver	
Battery	Rechargeable nickel-cadmium, 6 hours use from one charge
Dimensions	575 × 575 × 31-150mm
Weight	6Kg

Source

Swedish Geological Co.	Telephone +46 920 60300
Box 801	Telex 2401 8495029
S-95128 Lulea	GEONORD
Sweden	

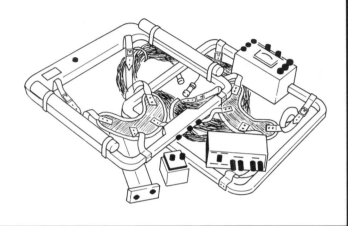

INDUCED POLARIZATION EQUIPMENT

A transmitter and its associated receiver together form the set of equipment needed for induced polarization geophysical surveying.

The transmitter is carried by backpack. It can be powered by batteries (disposable or rechargeable), or with AC power from this manufacturer's motor generator (described in Section 7). The transmitter has three available timing boards.

The receiver is a hand held unit with the ability to select positive or negative dc input, or a range of frequencies (0.125, 0.25, 1.0, 2.0 and 4.0 Hz as standard). It is housed in a non-conductive, impact resistant case.

Operating information

Transmitter IPT-1

Power	Disposable dry cells, or rechargeable batteries (48 volts at 9 Ampere hours) or AC input from the specified motor generator
Dimensions	20 × 40 × 55 cm
Weight	13 Kg with rechargeable batteries

Receiver IPV-1SP (illustrated)

Batteries	12 volt
Temperature	From minus 40 to plus 60 degrees Celsius
Dimensions	10 × 13 × 22 cm
Weight	1.1 Kg with batteries

Source

Phoenix Geophysics Ltd	Telephone 416 477 8588
7100 Warden Avenue	Telex 06-986856
Unit 7	
Markham	
Ontario	
L3R 5M7	
Canada	

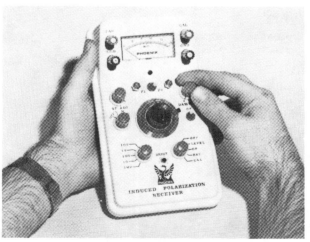

The IPV-1SP receiver

RESISTIVITY SURVEYING TRANSMITTER

A resistivity transmitter forms part of a set of equipment necessary for resistivity geological surveys.

The unit described here can be powered either from a rechargeable battery or with an AC power module (in which case a motor generator will also be needed). It is lightweight and highly portable, housed in impact resistant and non-conducting plastic, and can be carried in a rucksack.

The output voltage is adjustable in switchable steps, and its frequency can also be switched between two standard options (0.25 or 1 Hz). There is overload protection on the output current.

Operating information

Output voltage	53, 106, 212, 425 and 850V, switchable
Output current	3mA to 3 Amp
Output power	300 W, maximum
Output frequency	0.25 or 1Hz
Frequency stability	Plus or minus 0.005 per cent
Operating temperature	0 to 60 degrees Celsius, with 30 per cent of full capacity available at minus 40 degrees Celsius
Ammeter ranges	30mA, 100mA, 300mA, 1 Amp, 3 Amp and 10 Amp
Dimensions	20 × 40 × 55cm
Weight	13 Kg (17 Kg with AC power module)

Source

Phoenix Geophysics Ltd
7100 Warden Avenue
Unit 7
Markham
Ontario
L3R 5M7
Canada

Telephone 416 477 8588
Telex 06-986856

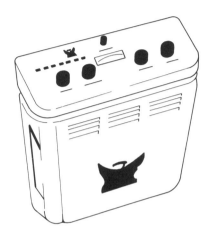

RESISTIVITY SURVEYING RECEIVER

Source

Phoenix Geophysics Ltd
7100 Warden Avenue
Unit 7
Markham
Ontario
L3R 5M7
Canada

Telephone 416 477 8588
Telex 06-986856

The resistivity receiver complements the resistivity transmitter from the same manufacturer (shown above) as part of the system required for resistivity geophysical surveying. Like the transmitter and its optional power generator, the receiver is designed to be easily portable. It is housed in a non-conductive, impact resistant case.

The unit is highly sensitive, but has a high tolerance to natural and other extraneous electrical noise. The input is protected against excessive voltages.

Operating information

Frequency range	0.25 to 5 Hz, selected in steps with an internal switch
Operating temperature	From minus 40 to plus 60 degrees Celsius
Temperature drift	Better than 2 per cent over the entire range
Display	The voltage can be read to 0.5 per cent of full scale
Calibration	An internal high stability 0.05 ohm resistor allows precise calibration under all conditions
Dimensions	10 × 13 × 22 cm
Weight	1.1 Kg

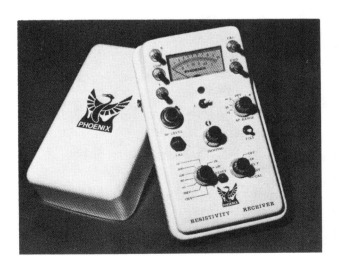

ELECTROMAGNETIC UNIT — VLF

This hand held instrument is used for geophysical mineral surveying. It requires no transmitter and is ideal for one man operation.

The unit measures the orientation and magnitude of the major and minor axes of the ellipse of polarization. These measured parameters are established using existing military and time standard VLF (very low frequency) transmitters. These are distributed throughout the world (see list on page 7) but, should there be no receivable station, a local electromagnetic transmitter can be used.

Two independent channels can be selected over a frequency range from 14 to 29.9 KHz, variable in increments of 100 Hz. The locking clinometer memorizes the tilt. A loudspeaker is provided for null indication.

Operating information

Battery	9 volt
Battery life	1 to 3 months, at a drain of 3mA
Temperature	Minus 40 to plus 60 degrees Celsius
Dimensions	8 × 22 × 14 cm
Weight	850 gm

Source

Phoenix Geophysics Ltd   Telephone 416 477 8588
7100 Warden Avenue   Telex   06-986856
Unit 7
Markham
Ontario
L3R 5M7
Canada

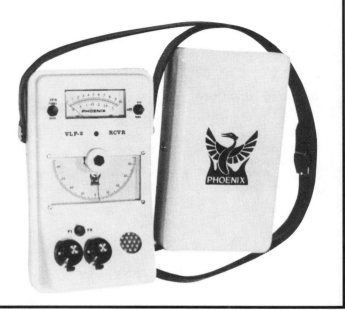

LIGHTWEIGHT PORTABLE SAMPLING DRILL

Simple in construction, light and easily portable, this drill is designed for drilling and sampling soil and rock.

Apart from its use in geological sampling, it can also be used as a general purpose drill for making holes in concrete for posts etc., during surface development and construction.

This drill is particularly suitable for use in inaccessible regions.

Operating information

Power	Single cylinder two-stroke 3 HP engine, running on a petrol/oil mixture
Drilling capacity	25m
Hole diameters	36mm, 46mm and 56mm
Drill rod diameter	24mm
Spindle speed	210 rpm
Feed	Hand wheel, giving 1000 Kgf maximum
Feed length	1m
Hoist	Hand winch, 100Kg capacity
Water pump	Two cylinder hand pump or mechanical
Mast height	3.5m maximum
Mast weight	75Kg
Drill unit dimensions	380 × 400 × 600mm
Drill weight	19Kg including power unit

This drill is one example of a range of drills and other equipment available from this source.

Source

China Geological   Telephone 668741
  Machinery and   Telex   22531
  Instrument Corp.
64 Funei Street Xisi
Beijing
People's Republic of China

WADI VLF INSTRUMENT

This is another instrument employing the VLF principle described on the previous page. The manufacturers of this unit describe its water finding use as locating structures where useful quantities of underground water may be trapped in rock fractures and cavities, thus indicating promising sites for well drilling. The results do not guarantee the presence of water; the conductive structure located might instead be an ore body for example. However, a good track record is claimed.

This unit has an LCD readout for displaying coordinates and graphical results. The readout will also state which VLF frequency the unit has locked on to, and there are versions to suit most languages. The complete system weighs 6Kg net, 1.6Kg of which is the hand held unit.

Operating information

Maker's type name	WADI
Power	Internal 9 volt dry cells (rechargeable option if required)
Battery life	About 2 weeks of field work
Station selection	Automatic
Frequency range	15 to 30 kHz in 100 Hz steps

Some VLF transmitters
VLF stations that can be used with the WADI include:

	Frequency (kHz)	Power (kW)
Bordeaux, France (FUO)	15.1	500
Rugby, Great Britain (GBR)	16.0	750
Hegeland, Norway (JXZ)	16.4	350
Gorki, USSR (ROR)	17.0	315
Moscow, USSR (UMS)	17.1	1000
Yosamai, Japan (NDT)	17.4	50
Oxford, Great Britain (GBZ)	19.6	

	Frequency (kHz)	Power (kW)
Annapolis, Maryland, USA (NSS)	21.4	400
Northwest Cape, Australia (NWC)	22.3	1000
Laulualei, Hawaii, USA (NPM)	23.4	600
Buenos Aires, Argentina (LPZ)	23.6	
Cutler, Maine, USA (NAA)	24.0	1000
Seattle, Washington, USA (NLK)	24.8	125
Aguada, Puerto Rico (NAU)	28.5	100

Display	Easy-to-read LCD, 150 × 40mm
Inclinometer range	Minus 10 to plus 10 degrees, 0.2 per cent precision
Maximum sensitivity	Better than 100 nanoamp per metre
Minimum sensitivity	500 milliamp per metre
Transmitter range	Between about 75 and 10,000 km from a powerful transmitter (see table)
Memory capacity	Data from about 6,000 measuring stations
Data output	Serial RS232C to separate dot matrix printer (optional accessory)

Source

Atlas Copco	Telephone 08 764 60 60
ABEM AB	Telex 13079
Group Centre for Geophysics and Electronics	
Box 20086	
S-161 20 Bromma	
Sweden	

Please note that this manufacturer has supplied information on a range of lightweight geophysical electronic instruments for prospecting and surveying, all of which use modern technology.

TWO LIGHT DRILLS

Conrad-Banka Drill

This is a light hand-operated drill rig which is suitable for use in remote areas for light civil engineering tasks and in prospecting for:

— Tin ores, gold, platinum, ilmenite, rutile, monzite and other minerals in alluvial deposits
— Alluvial ore deposits and tailings
— Phosphates, china clay, bauxite and lateritic iron ore.

The drill will penetrate shale, soft slate, hard cemented gravel and weathered decomposed bedrock. It is not suitable for drilling in hard rock.

Note: This drill has been widely used for almost 100 years, and models are produced locally in many countries.

Power Pioneer (PP150 — Mk 2) and Banka mechanized drill

This is a semi-mechanized percussion rig which uses the same casings and most of the drills applicable to the Conrad-Banka drill. Here, however, the casing is not lowered by rotation, but mainly by ramming. The upward and downward movement of the drilling tools, which are suspended by wire rope, is effected directly by the chain-driven hoisting drum, which is provided with a friction clutch and friction brake. The winch motor can be either a diesel or a petrol engine.

This rig is used for waterwell drilling, site investigation and investigation of alluvial ore deposits. The drill can reach a depth of about 45 metres using 4 inch, 6 inch and 8 inch casings.

As an option, use can be made of a hydraulic rotary table operating at 2 rpm (500 Kg for the 4 inch and/or 6 inch casing and 1000 Kg for the 6 inch and/or 8 inch casing). Power is taken through a telescopic torque arm and a hydraulic unit powered from the winch engine.

Sources

Arkana Europe BV   Telephone 05270-16965/14141
Produktieweg 15
PO Box 1004
8300 BA
Emmeloord
Netherlands

Conrad-Stork   Telephone 023-319170
PO Box 134   Telex   41048
Haarlem
Netherlands

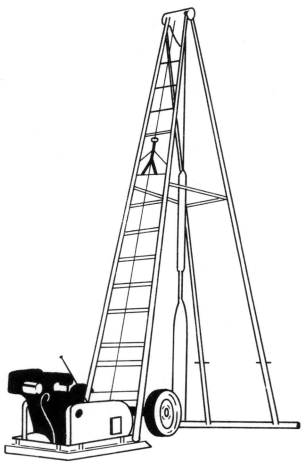

Conrad-Banka Drill

PORTABLE DIAMOND DRILL

Two men can carry this lightweight and compact drill, or it can be airlifted by helicopter. The drive is powered by a 10 HP, air-cooled, two-stroke petrol engine and the drill can be operated at any angle and at high altitudes.

The drill can be hand held, 'Packsack', or it can be mounted on a press (in which case the pressure is exerted via a hand wheel or 'winkie'). The drill is primarily set up for diamond drilling and, using lightweight drill steels, the following depths can be achieved.
— EX barrel, 22mm core to a depth of 120 to 138m
— BX barrel, 41mm core to a depth of 45 to 60m.
By attaching an auger reduction unit, drilling can be carried out to a depth of 30.5m with a 66.7mm auger.

Operating information

Power	Two-stroke, air-cooled, petrol engine producing 10 HP at 8400 rpm
Clutch type	Centrifugal
Max. bit speeds	2800 rpm in high gear, 1200 rpm in low gear
Dimensions	530 × 530mm
Weight	74.8Kg

Source

JKS Boyles Ltd	Telephone 0623 754482
Byron Avenue	Telex 377994
Lowmoor Industrial Estate	
Kirkby-in-Ashfield	
Notts	
NG17 7LA	
UK	

MOBILE SHALLOW DRILL

This is a very lightweight mobile drill which is useful for shallow depth mineral prospecting. The power from the petrol engine is transmitted to the drill through a centrifugal clutch and sealed gearbox. The drive arrangement allows high speed/low torque and low speed/high torque combinations to be used. There are several operating variants.

The auger kit option allows 76.2mm holes to be drilled to a depth of 9.14m. This kit is useful for shallow exploration sampling.

A soil sampling kit allows penetration testing down to 15.2m. Core drilling to the same depth can also be undertaken, with a core diameter of 22.2mm.

Operating information

Power	Briggs and Stratton four-stroke petrol engine, 7 HP, air-cooled
Torque	48 Kgm maximum at the spindle
Transmission	3 forward speeds and 1 reverse
Dimensions	584 × 559 × 1613mm
Weight	106.6 Kg

Source

Mobile Drill International Inc.	Telephone 317 787 6371
3807 Madison Avenue	Telex 27352
Indianapolis	
IN 46227	
USA	

PERCUSSION DRILL RIG

A diesel engine powers this medium-sized percussion cable drill. It can drill 457mm diameter holes to a depth of 389m, and is suitable for water wells, blast holes, seismograph and prospecting holes.

The unit is mounted on pneumatic tyres for towing. Alternatively, a skid-mounted version is available for truck haulage. The collapsible derrick allows one-man operation.

Drill hoist and winches are all driven from the diesel engine. Bull and casing wheels are gear driven, whilst the sand wheel is powered through a friction drive.

Operating information

Engine	Diesel, 50 HP, fitted according to requirements
Derrick height	12.19m
Hoist speed	182m per minute
Weight	5.8 T, without cable or tools
Travelling dimensions	
— Height	3.05m
— Length	8.2m
— Width	2.19m

Source

Southern Drilling Supplies Ltd	Telephone 0491 571181
Reading Road	Telex 847597
Henley-on-Thames	
Oxon	
RG9 1DX	
UK	

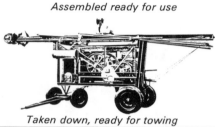

Assembled ready for use

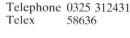

Taken down, ready for towing

MOBILE ROTARY DRILL

This is a top-drive multi-purpose lightweight mobile drill. It can be supplied with transport, but the basic unit can be mounted on a variety of wheeled or tracked carriers.

Drills can be built to customers' requirements. They can incorporate inclined drilling, free fall winch and mast extension (allowing 6m pipe lengths to be handled).

The rig can be used to achieve the following capacities:
— Continuous flight auger, 152mm diameter, 30m
— Down-the-hole or rotary, 203mm diameter, 114mm pipe, 213m
— Down-the-hole or rotary, 152mm diameter, 89mm pipe, 305m
— Reverse circulation, 610mm diameter, 168mm pipe, 90m
— Wire line coring, HQ rod, 366m
— Wire line coring, NQ rod, 550m

Operating information

Rotary head output (standard model)	0 to 160 rpm and 4000 Nm maximum torque
Pull down system:	
— Pulldown	5000 Kgf
— Pullup	8000 Kgf
— Speed down	58m per minute maximum
— Speed up	30m per minute maximum
— Stroke	3.9m
Mast:	
— Safeworking load	8000 Kgf
— Lifting clearance	5m
Power unit	Deutz, air cooled, 41.7 Kw at 2000 rpm

Source

UMM Ltd	Telephone 0325 312431
Drilling Division	Telex 58636
PO Box 9	
Aycliffe Industrial Estate	
Newton Aycliffe	
Co. Durham	
DL5 6DS	
UK	

CORE DRILL

This portable drill is designed for shallow well drilling. It is applicable to general investigation, geophysical exploration, blast hole drilling and construction work.

It is lightweight and easily demountable for transportation (it can break down into ten separate parts if required).

Handwheel or powered drill feed can be used, to suit different drilling requirements.

The primary power source required is a 12 HP diesel engine, with a speed rating of 1800 rpm, which can be supplied (as illustrated).

Operating information

Drilling depth	100m
Initial hole diameter	110mm
Final hole diameter	75mm
Drilling pipe diameter	42mm
Hoist:	
Lifting capacity	1000 Kgf maximum, one speed, single rope
Rope	9.3mm wire
Drum capacity	27mm of wire rope
Water pump:	
Type	Horizontal double acting single cylinder
Maximum displacement	80 litres per minute
Working pressure	7 Kgf/cm^2 (15 Kgf/cm^2 maximum pressure)
Suction pipe	38mm diameter
Displacement pipe	32mm diameter
Power required	3 HP

Source
China Geological Machinery and Instrument Corp.
64 Funei Street Xisi
Beijing
People's Republic of China

Telephone 668741
Telex 22531

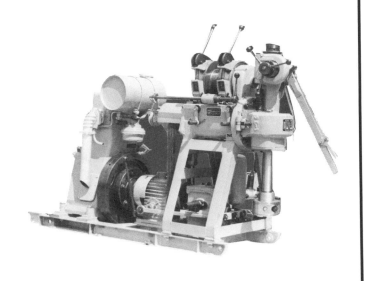

THEODOLITE

The manufacturer describes this instrument as a transit vernier theodolite, 20 seconds, of proven performance, robust but light construction and supported on a telescopic or rigid stand with press-down feet.

The theodolite is supplied in a weatherproof varnished wooden case, along with standard accessories.

Operating information

Type	Th-2
Telescope length	178mm
Aperture	40mm
Magnification	28 ×
Nearest focus	2.3m
Resolution	300m
Stadia ratio	1:100
Horizontal circle	125mm diameter
Vertical circle	114mm diameter
Vernier reading	20 seconds
Plate bubble	30 seconds
Altitude bubble	20 seconds

This manufacturer offers many other models of levels, theodolites and accessories.

Source
East End Engineers
52 Bhupendra Bose Avenue
Calcutta-700 004
India

Telephone 55-9493

COMBINED LEVEL AND COMPASS

This is a solidly built instrument with a slow motion adjustment. The instrument is supplied in a wooden case, complete with rigid stand and standard accessories.

Operating information

Type	DI-12C
Telescope length	305mm
Aperture	37mm
Magnification	20 ×
Nearest focus	3m
Resolution	200m
Stadia ratio	1:100
Horizontal circle	Nil
Minimum angle reading	Nil
Spirit level	20 seconds
Compass dial	72mm diameter, centrally mounted
Compass divisions	30 minutes (prismatic reading arrangement)

Source

East End Engineers   Telephone 55-9493
52 Bhupendra Bose Avenue
Calcutta-700 004
India

LEVELLING INSTRUMENTS

We have received details, but are not able to illustrate, two items of levelling equipment as described below. Both are available from the same source, listed at the foot of this entry.

Abney level

This is a simple hand held level costing about $53 US (1989 price). Its radius arc has two separately graduated scales marked in degrees (0 to 90 degrees in both directions). A vernier scale allows a reading accuracy of 10 seconds of arc. The radius arc is also calibrated as a percentage scale, from zero to 100 per cent. A leather case and belt loop are provided. This instrument weighs 0.25Kg and its length is 159mm, extending to 178mm.

Levelling staff

A levelling staff is used in conjunction with a levelling instrument for determining levels. The staff described here is graduated in feet, tenths and hundredths. It is telescopic, or can be broken down into four separate sections. The material is fibreglass, coated with epoxy resin.

A levelling bubble is provided, and a carrying case is also supplied. The length of this staff is 1.7m extendable to 4.3m. It costs approximately $135 US (1989 price).

Source

Miners Inc.   Telephone 208 628 3247
PO Box 1301   Telex   150030
Riggins
Idaho 83549
USA

PROGRAMMABLE MINERAL ANALYSER

This device uses X-ray fluorescence or X-ray transmission for low cost assays of metals in samples. Radioactive isotopes are used as radiation sources. The illustration shows the X-ray fluorescence head in use. The instrument analyses the pulse emission or absorption pattern of samples, each element producing its own unique and identifiable pattern. The system comprises:
— Analysis head for X-ray fluorescence
— Analysis head for X-ray transmission
— Electronics unit.
Depending on which head is used, the system can measure elements such as tin, iron, nickel, copper, zinc, molybdenum, antimony, lead, bismuth, tungsten, tantalum and uranium. Alloys and some other materials can also be analysed. Samples can be in the form of metallurgical mill products (pulverized) or solutions.

A microprocessor is used to control the setting up and measurement procedures. Operation is claimed to be simple and the supplier can train the user.

Assays are automatic, and normally take two minutes. The results are printed out on the built-in dot matrix printer.

Operating information
It is recommended that the system is used in a dust-free, constant temperature environment if possible. In any case the operating temperature must not exceed 40 degrees Celsius.

Analysis head details:	*Fluorescence*	*Transmission*
Bench area	250 × 200mm	220 × 190mm
Height	300mm	400mm
Weight	8 Kg	9 Kg

Electronics unit details:

CPU	8085A
Memories	8 × 32K EPROMs and 8 × 3K RAM with battery back-up
High voltage	500 to 1500 volt, programmable in 1 volt steps for automatic gain stabilization
Pulse height discriminators	Six channel analysers with programmable discriminator levels
Printer	Built in dot matrix, 18 characters per line
Displays	12-character alphanumeric LED displays 10 LED status lights
Power input	200-260 volts or 100-130 volts, 50 or 60Hz
Dimensions	350 × 500 × 170mm
Weight	12 Kg

Source
Amdel Ltd   Telephone 08 372 2700
Instrumentation Division Telex   AA 82520
31 Flemington Street
Frewville
South Australia 5063

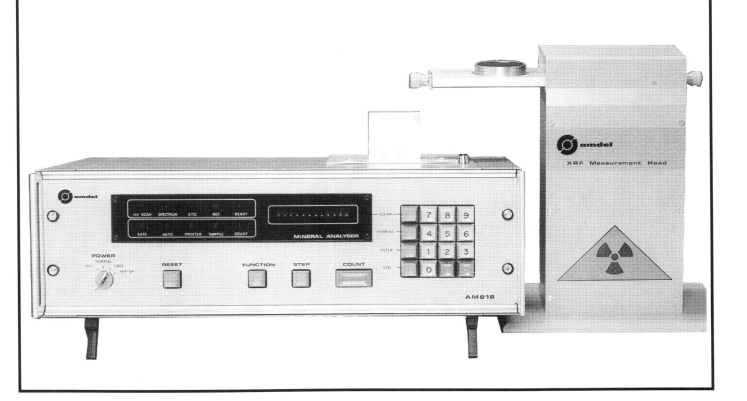

GEOLOGICAL STRATUM COMPASS

A compass for geological, physical and mining applications. Azimuth and angle of dip can be measured.

One side of the compass has a 60mm graduated ruler and the hinge has a marked vertical circle. In the centre of the compass there is a levelling bubble, which is illuminated by a mirror on the underside.

The circular magnet is provided with eddy current damping and the compass needle can be locked by a lever.

It should be noted that this particular example is a more complex, and therefore expensive, type of compass.

Operating information

Vertical graduation	5 degrees
Reading accuracy	Plus or minus one degree
Horizontal circle graduation	1 degree
Reading accuracy	Better than 0.1 degree
Dimensions (closed)	73 × 95 × 25mm
Weight	250g

Source

Breithaupt Kassel   Telephone 208 628 3247
c/o Miners Inc.   Telex   150030
PO Box 1301
Riggins
Idaho 83549
USA

or direct from manufacturer in West Germany.

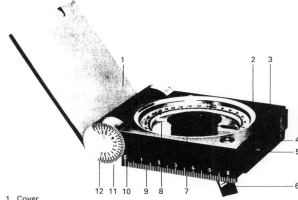

1. Cover
2. Compass housing
3. Spring pin front sight
4. Drive screw for adjusting the circle to the local declination
5. Eye for the carrying cord
6. Illumination mirror for round spirit level
7. Damping pot
8. Magnetic needle
9. Tracing edge
10. Index line
11. Hinge adjustment
12. Vertical circle

2. Surface development and working

This section covers the equipment utilized in the opening up and working of small surface mineral deposits. The types of deposits envisaged range from coal, hard rock minerals, soft sandy types, to deposits found under water.

Some types of equipment used in surface workings are covered in Section Five and so will be mentioned only briefly in this introduction.

In discussing the equipment, it is presumed that the exploratory work and site clearing has been completed and that roads to the work site are suitable for the equipment to be used.

In some cases the equipment can be moved to site under its own power, in some the equipment will be transported intact and in other cases it will need to be assembled on site.

In this section only a small sample of the very large range of equipment available in the world market is illustrated; because of the large selection available it is advisable when purchasing any item of equipment to seek assistance from the manufacturer or supplier. With this in mind, a number of additional manufacturers' addresses have been supplied.

Drills and rock breakers

Hand-held

Hand-held drills, which can be operated by compressed air or petrol driven, are used to drill small diameter holes prior to loading with explosive to assist in breaking up the hard rock.

Softer material can be excavated by means of special attachments, avoiding the use of explosives, or by other means of breaking rock.

Drill rigs

Larger drills are available for use with crawler- or truck-mounted drilling rigs, and these may be compressed air or hydraulically operated. The drill can be mounted on the rig or operated at the end of the drill string as a down-the-hole hammer (DTH).

As shown, these drill rigs can be permanent or a specialized attachment for when only intermittent use is envisaged, allowing one unit of equipment, e.g. an excavator, to operate in several ways.

Drill bits and drill steels

Drill bits are available in the form of tungsten carbide button bits or as solid tungsten carbide inserts. Bits larger than 50mm suitable for use with drill rigs are always detachable by means of a thread or taper fit from the drill rod or tube. Bits smaller than 50mm are generally integrated with the drill steel. Both bits require a device for sharpening if the full life possible is to be obtained from the tungsten carbide (T/C) inserts.

Drill steels come in various sizes depending on the duty to be performed. Consultation of suppliers' catalogues is necessary to determine the correct size to be used. Drill rigs can use solid steels or drill tubes depending upon the type of drill used. DTH drills require tubes.

Rock breakers

Rock-breaking tools are available in various forms, the most common being the hydraulically operated version mounted on an excavator boom and used in quarries and mines in place of secondary blasting. They are also used to demolish buildings and to break solid rock where blasting is not possible. Other versions illustrated are used to maintain grizzleys open and to clear jams at crushers.

If explosives are not available or permitted, but breaking of rock is required, rock-splitting attachments can be used with hand-held drills.

The attachments can be manually or hydraulically operated.

Explosives

Explosives in various forms are widely available from many manufacturers ranging from cheap ammonium nitrate based explosives to highly sophisticated explosives used in oil well technology.

There are many ways available of detonating the explosive and expert advice should be sought if the use of explosives is contemplated. Also, there are invariably local rules and regulations regarding use and storage; these must always be studied and adhered to.

Excavators

Excavators come in many forms and sizes, from small $\frac{1}{4}$m³ capacity to 50m³ and higher. There are several types illustrated here that generally have the same purpose, which is to dig holes in the ground or to load broken material on to a means of transport. However, they are versatile, with many attachments available to enable different duties to be undertaken with the same basic unit. Some are more mobile than others; some are electrically operated and others use diesel-generated power. Some excavators are tracked and others rubber tyred depending upon the degree of mobility required.

In general, backhoe-type excavators are tracked and front-end shovel types are wheeled.

As excavators can be used for many purposes, consultation with a manufacturer is always advised if a unit is required to carry out an alternative duty.

Water monitors

These are a means of removing unconsolidated soft, sandy, clayey and gravelly material by use of a high-pressure water jet, usually employed in conjunction with an electric or diesel pump and adequate water supplies.

The stream of water is used to carry the material to a sump where gravel or sand pumps collect the material and pump it to the treatment plant.

Dredges

These are operated on the open sea close inshore, on lakes, on rivers, or on specially constructed ponds. Excavation of the underwater material can be by means of backhoe, clam shell, bucket ladder, cutter suction head, suction pumps or bucketwheel.

Dredges come in many sizes with the larger sizes designed specifically for the intended operation. Even the smallest machines are relatively complex and should therefore be considered for use only by adequately trained persons.

Alan Baxter

LIGHTWEIGHT COMBINATION DRILL AND BREAKER

Needing no connecting pipes, this small drill/breaker is self powered by a built-in petrol engine. It is light enough to be carried to remote or high places on a back pack.

Automatic throttle control is fitted to the model described, which is actuated by depressing the anti-vibration drill handles. When pressure on the handles is released, the drill reverts to idling speed.

A comprehensive range of tools is available, and a power take off allows tools to be reground on the spot. It can be used to pump water.

The unit is supplied in a protective case (see illustration) and comes with maintenance tools, drill steel and concrete chisel as standard.

Source
Berema AB
Box 1286
S-17125 SOLNA
Sweden

Operating information

Type	Cobra 148ATC*
Power	128 cc air-cooled single cylinder two-stroke engine running on a petrol/oil mixture
Ignition	Thyristor type
Fuel consumption	1.1 to 1.5 litres per hour
Tank capacity	1.6 litres
Drilling depth	6m maximum (can be used horizontally)
Drilling rate	250 to 300mm per minute in granite with 34mm bit
Weight	24 Kg

*An alternative model is available for breaking use only.

DRILL RIG ATTACHMENT FOR EXCAVATOR

This drill rig fits on to the dipper arm of a standard excavator (with the bucket removed). The control panel is mounted next to the drill.

Provided the hydraulic system and mountings are suitable, the drill can be fitted to any excavator. Smalley excavators are especially designed to accept this equipment.

The combined excavator and drill is suitable for sites where space is limited and where there is scope for standard units to be used for multirole purposes. In small operations the same unit can be used to drill blast holes and load the broken material after blasting.

Source
Boart UK
Littlemoor
Eckington
Sheffield
S31 9EF
UK

Operating information

Rock drill	Boart HD-65
Feed (rock drill travel)	2.16m
Bits	48mm button bit (other options available)
Hydraulic requirements:	
— Percussion	65 litres per minute at 150 bar Feed thrust, 10 KN
— Rotation	15 litres per minute at 80 bar Rotation torque 80 Nm
— Feed	4 litres per minute at 80 bar
Power requirement	25 kW

DRILLING ATTACHMENT FOR LOAD-HAUL-DUMP BUCKET

This special drill rig/drifter attachment is designed to fit the LHD bucket manufactured by the same company. Two versions are available, horizontal and vertical.

The drill is attached to the bucket by just sliding it over the front lip and then connecting the hydraulics.

Although the effect is not to produce a purpose-built drill rig, it does provide a flexible and initially cheaper option than the purchase of two separate units.

Operating information

Stroke	2m
Feeder length	2.8m
Height	2 to 4m, depending on the loader to which the unit is mounted

Unit dimensions:

— Height	1.65m
— Length	3.00m
— Width	0.75m
— Weight	460 Kg

Source

France Loader   Telephone (1) 45 01 81 24
50 Avenue Victor Hugo   Telex   610713
75116 Paris
France

BREAKER

This is an electrically powered hydraulic breaker, designed for simple installation wherever a stationary breaker is needed. There is a range of hammers to choose from.

Operating information

Power	30 or 40 HP electric motor, 415 volts, 3 phase
Power pack	Skid mounted
Power pack weight	550 Kg

Oil reservoir	40 gallon, in the power pack
Minimum reach	1m
Maximum reach	4m
Maximum height reached	2.4m at 3m reach
Maximum depth reached	4.6m at 1.6m reach

Source

Smalley Excavators Ltd   Telephone 0778 426426
Cherry Holt Road   Telex   32225
Bourne
Lincs
UK

The manufacturer is Krupp of West Germany.

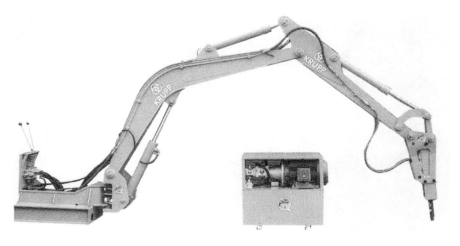

PNEUMATIC CRAWLER DRILL

This tracked vehicle can be purchased in several configurations, which can be used for a variety of drilling work including quarrying and open pit mining. Towable compressors suitable for providing the pneumatic power can also be supplied by Atlas Copco. The traction and winch controls are mounted on a swivelling arm, for safety and ease of operation.

The version described here can be used for bench drilling, long-hole drilling, overburden drilling or water well drilling.

Operating information

Crawler details:

Type	ROC 400A-00
Travel speed	2.5 km per hour maximum
Tractive pull	40 kN maximum
Ground clearance	350mm
Hill climbing	30 degrees maximum
Ground pressure	0.06 N/mm^2
Chain feed travel	3.9m
Feed force	12.3 kN
Feed travel speed	0.42 metres per second

Rock drill attachment:

Type	BBE 57-01 for the tasks described above
Stroke	66mm
Impact rate	2000 blows per minute
Rock drill tools	Sandvik Coromant range
Drill bit diameter	64 to 115mm
Air requirement	374 litres per second (maximum) at 6 bar for this crawler/drill combination

Suitable compressor	Atlas Copco type XAS430
Weight	Approximately 5000 Kg, depending on equipment fitted

Source

Atlas Copco
 (Great Britain) Ltd
PO Box 79
Swallowdale Lane
Hemel Hempstead
Herts
HP2 7HA
UK

Telephone 0442 61201
Telex   825963

41241-100

DRILLING EQUIPMENT SUPPLIERS

The following list of manufacturers/suppliers specialize in the various types of drills and drilling equipment. Many are listed elsewhere in the Guide, but have been placed together here under the heading of drilling for ease of reference.

Hand-held drills
Compair Holman Ltd   Telephone 0209 712750
Camborne
Cornwall
UK

Drills for drill rigs
Atlas Copco Mct AB   Telephone 8-7438000
S-10484
Stockholm
Sweden

Ingersoll Rand Co.   Telephone 201 573
200 Chestnut Ridge Road
Woodcliff Lake
NJ 07675
USA

Down-the-hole hammers
Weaver & Hurt Ltd   Telephone 0246 450608
Station Lane
Old Whittington
Chesterfield
Derbyshire
UK

Compair Holman Ltd   Telephone 0209 712750
(as above)

Rock splitters
Compair Holman Ltd   Telephone 0209 712750
(as above)

Atlas Copco Mct AB   Telephone 8-7438000
(as above)

Rock breakers
Montabert   Telephone 78 90 8122
203 Route Grenoble
BP 671
F-69805 St. Priest Cedex
France

Kent Air Tool Co.   Telephone 216 673 5826
71 Lake Street
Kent
Ontario
Canada

Tracked and tyred drill rigs
Compair Holman Ltd   Telephone 0209 712750
(as above)

Tamrock   Telephone 358 31 241 411
SF-33310
Tampere
Finland

Drill bits and drill steels
Boart UK Ltd   Telephone 0628 75311
39 Queen Street
Maidenhead
Berkshire
UK

Sandrock Rock Tools   Telephone 26 260000
S-81181 Sandviken
Sweden

MICRO-EXCAVATOR

Constructed as a trailer for towing by a vehicle, this small micro-excavator is highly mobile. It is also extremely simple in design, with low maintenance requirements.

Machines such as this are useful for all manner of small excavations, especially in the mechanization of placer mining.

The excavator has a low centre of gravity and widely spaced anchor points, making it stable and safe in use.

Operating information
Bucket width	200m up to 300mm are available

Power	Honda air-cooled four-stroke single cylinder petrol engine or a diesel alternative
Pump	105 Kg per cm²
Breakout force	1.9 tonne
Hydraulic tank	18 litres capacity
Length	3.05m
Height	1.7m
Width	1.68m

Source
Team Services International Ltd	Telephone 0536 711246
Kyngswoode	Telex 341 543
Rushton Road	
Rothwell	
Northants	
UK	

MINI-EXCAVATOR

This small excavator, for light materials, is versatile and easily transferable from site to site. The excavator is equipped with steel crawler tracks, which result in the low ground bearing pressure of only 0.23 Kgf per cm².

Mini-excavators such as this have a proven record of high reliability with low maintenance requirements.

Additional options are available. Particularly interesting is the hydraulic breaker, which is useful for secondary breakage after poor fragmentation from a blast.

Before choosing a mini-excavator for placer mining, careful account should be taken of the breakout force necessary to fragment the gravel.

Operating information
Engine	4 cylinder diesel, 33 HP
Grade ability	30 degrees
Breakout force	2500 Kgf
Travel speed	1.8 to 3.3 kilometres per hour
Safe working load	380 Kg at full reach
Fuel consumption	4.89 litres per hour at 60 per cent loading
Range of buckets	457 to 762mm

Source
Kubota (UK) Ltd
Dormer Road
Thame
Oxon
OX9 3UN
UK

TRACTOR-MOUNTED BACKHOE

A backhoe can be used for a variety of handling and excavating tasks. This version is designed for attaching to any tractor with three-point linkage and hydraulic power take-off. With a bucket breakout force of 1830 Kg, it would be suitable for light excavations during placer mining.

The device has stabilizers and a seat for the operator. Its control unit consists of five spool valves, operating five hydraulic cylinders.

Operating information

Depth of dig	2.5m
Height of dig	2.9m
Maximum reach	3.1m
Pulling force of bucket arm	1125 Kg

Load capacity	200 Kg at maximum extension
Operating sweep	180 degrees
Hydraulic system:	
Flow	10 litres per minute, minimum
Pressure	135 bar

This information relates to the Scandig 60T unit. A smaller version, the 40T, is among the range of equipment available from this company.

Source

Scan Tonga Engineering Co. Ltd
Private bag
Nuku'alofa
Kingdom of Tonga
SW Pacific

Telephone 676 22599
Telex 66203 TONENG TS

Other sources

JCB Sales Ltd
Rocester
Staffs
ST14 5JP
UK

Telephone 0899 590312
Telex 36372

SMALL EXCAVATOR

This small tracked excavator is available with or without a cab. It can dig in an offset position, which allows working within 20cm of a wall or other obstruction.

The hydraulic system has three pumps, one for digging and slewing and one for each track. The hydraulics are powered by a Lister two cylinder air-cooled diesel engine.

Small excavators of this type are used worldwide for a big range of loading and excavating tasks. Open cast mining and the mechanization of alluvial placer mining are only two of the many suitable operations.

This excavator can be fitted with a variety of additional tools.

Operating information

Capacity	0.115m³
Digging depth	2.45m (2.7m with the machine tilted)
Reach	4.4m
Tear out force	2333 KgF at bucket
Gradeability	37 degrees
Ground pressure	0.27 Kg/cm
Working speed	1.67 km per hour
Travel speed	3.7 km per hour
Power	Lister diesel, 20.08 brake horse power
Weight	2.8 tonnes

Source

Smalley Excavators Ltd
Cherry Holt Road
Bourne
Lincs
UK

Telephone 0778 426426
Telex 32225

DYNAMO BLASTING MACHINE

The bare ends of detonator wires are connected to the terminals of this device after continuity testing the detonator circuit. The firing voltage is generated by a dynamo within the device, which is turned by hand.

The machine is grasped in the left hand, and the right hand holds the firing handle. The handle is turned sharply clockwise, and the body is simultaneously turned in the other direction until rotation is halted by an end stop (amounting in total to about one third of a turn).

At the end of the drive movement, when the highest rate of rotation (and, therefore, power generation) has been reached, the machine automatically applies the firing voltage to the detonators for a period of four milliseconds.

Operating information

Number of shots	25 maximum
Voltage	130 volts
Current	1.2 amps
Pulse duration	4 milliseconds
Limiting resistance	110 ohms

This unit is manufactured by IDL Chemicals Ltd under licence from the Austrian firm Schaffler and Co.

Source

IDL Chemicals Ltd   Telephone 260570
P.B. No 1   Telex   0155-6243
Sanatnagar (IE) P.O.
Hyderabad-500 018
A.P. India

Another company which supplied information on blasters, continuity testers and explosives for this Guide is:

M/S Industrial   Telephone 48631 or 48632
  Explosives (P) Ltd   Telex   0715-222
Maimoon Chambers
Central Avenue
Post Box 352
Gandhibagh
Nagpur-2
India

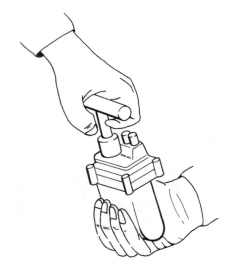

BLASTING METER

This is an example of an instrument which tests the continuity of the electric connections before firing. The testing current is low, and limited to only 4 mA even under short circuit conditions. Detonators cannot, therefore, be accidentally fired by the test current.

The meter is rugged, and there is no sensitive moving coil meter to be damaged, indication being by light emitting diodes (LEDs). It is capable of being used underground.

Operating information

Power	Two internal dry pencil torch cells
Meter range	There are three versions: 0 to 100 ohms or 0 to 500 ohms or 0 to 1000 ohms (The meter chosen will depend on size of the round)

Source

IDL Chemicals Ltd   Telephone 260570
P.B. No.1   Telex   0155-6243
Sanatnagar (IE) P.O.
Hyderabad-500 018
A.P. India

Explosives are available from:
Nitro Nobel AB   Telephone 587 85000
Gyttorp
S-71382 Nova
Sweden

Du Pont de Nemours & Co.
Wilmington
Delaware 19898
USA

CUTTERHEAD DREDGE

The cutterhead dredge described here is suitable for one man operation. It is portable, and can be loaded on to a road trailer.

The dredge is capable of handling solids up to 203mm diameter using a vortex pump. A variety of cutting heads are available, and the cutting boom itself can traverse through 90°.

The unit is floated by 18 separate modular pontoons and removable outriggers. Propulsion is by an outboard engine (2.5 to 20 horsepower). A hydraulic winch is used for securing and launching.

Such units are used worldwide for dredging.

Operating information

Engine   Chrysler 8 cylinder, 150
                          horsepower
Fuel consumption   5 to 8 gallons of petrol per
                          hour
Weight   4422 Kg
Length   6.5m
Width   2.2m
Draft   411mm

A range of models is available, with petrol or diesel engines. A gold and gem recovery sluice system can be supplied for use with this dredge.

Source

Keene Engineering Co.   Telephone 818 993 0411
9330 Corbin Avenue   Telex   662617
Northridge
California 91324
USA

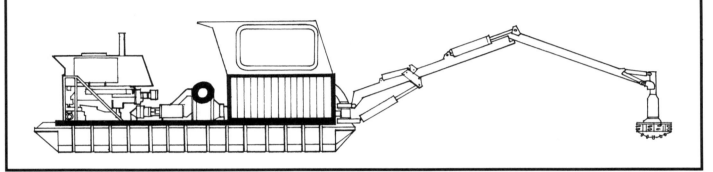

3. Underground development and working

This section covers some of the equipment which might be used in gaining access to the orebody, and then in mining the ore.

The mode of access depends on the size and shape of the orebody, its situation and the strength of the orebody and surrounding 'country' rock. Access can be gained by means of an adit, a horizontal drift into a hillside; a decline, a low angle inclined drift; or a shaft which may be vertical or inclined.

The access can be mined in the orebody itself, the cheapest way, or in the country rock outside the orebody. The choice will depend on the size of the orebody, and hence its life, the value of the orebody, and hence the capital available, and the relative strengths of the orebody and the surrounding rock.

Shafts

Steeply inclined and vertical shafts, unless shallow in depth and small in area, normally require the use of specialist equipment for mining them, hoisting the excavated rock and transporting men and materials. This work is best left to specialized contractors.

Underground development

Main development includes access declines and adits and all horizontal or near horizontal tunnels to reach the orebody.

Small tunnels are normally drilled using jackhammers mounted on airlegs. Larger tunnels could be drilled using electro-hydraulic or pneumatic mobile drill rigs, but these machines are very expensive.

Broken ground from the development would normally be removed using air-powered rocker shovels for loading and locomotives and mine cars for transportation or load-haul-dump vehicles for loading and trucks for transportation as described in Section 5.

The use of drill rigs and load-haul-dump vehicles entails the provision of workshops and a fairly high degree of engineering skills.

Stope development, normally in or very close to the orebody, includes drives, crosscuts and raises from which the orebody will be drilled and blasted. The drives and crosscuts will be drilled using jackhammers and airlegs or small jumbos and the raises using jackhammers and airlegs. Mucking of the broken rock would be by rocker shovel, load-haul-dump vehicle or perhaps by scraper.

Stoping (working, or mining of the orebody)

Following on from development of the orebody is the process of stoping, the drilling and blasting of the orebody. The method of stoping will depend on the width and dip of the orebody, its competence and that of the surrounding rocks.

For flat dipping orebodies, room and pillar stoping might well be used, using jackhammers and airlegs or mobile drill-rigs to drill the ore. For narrow, steeply dipping orebodies drill rigs could be used.

Support

In hard rock mining, most support is provided by roof bolts either mechanically fixed or grouted using resin or cement. Other methods use split set stabilizers which are forced into an undersize hole, or swellex bolts which are expanded into a hole by water pressure.

Timber is often still used in the form of props, sets or packs, particularly where good, cheap timber is available. Hydraulic props and chocks are expensive and heavy and are mainly used in highly mechanized coal mining.

Mine ventilation

Mechanical ventilation by means of fans is normally required in any mine to ensure a sufficient supply of oxygen and to remove dust and gases produced in mining operations.

The number and size of the fans required will depend on the size and complexity of the operations. The fans may be static, mounted in a housing on surface at the top of a shaft or may be movable and deliver or exhaust air through flexible or rigid columns. Most fan manufacturers will assist in designing ventilation circuits and will suggest the type and size of fans to be used.

Mine surveying

All of the development mentioned above is required to be surveyed and the information plotted on plans of the workings. This is normally a statutory requirement apart from being essential to the efficient operation of the mine. A trained surveyor will be needed for this job. His main tool is the theodolite which is used for measuring angles, both horizontal and vertical. Theodolites are normally used mounted on a tripod. Special varieties may be suspended from the roof in low workings with insufficient headroom.

Mine communications

Communications systems will vary according to the size and complexity of the mine. In small mines the only communications system is likely to be the signalling system between the hoist driver and the cagetender. Increasing complexity leads to telephone systems and then to radio systems, which may also be used for controlling equipment.

Jack Simpson

UNDERGROUND MINE DRILLING MACHINE

From the Jixi Coal Special Equipment Factory in China, this drill can be used in underground mines for drilling holes with various angles up to 200m deep. It can also be used on the surface for geological drilling tasks.

Operating information

Catalogue number	MAZ-200
Drilling depth	200m
Drill rod diameter	42 to 50mm
Diameter of end hole	75mm
Drilling spindle speed	80, 155, 215 or 410 rpm, with vertical spindle
Hole angle	0 to 360 degrees
Reverse speed	50 and 100 rpm
Spindle stroke	400mm
Feeding force	39.2 KN maximum
Hydraulic winch load	9.8 KN
Motor power	10 kW
Dimensions	145 × 900 × 1450mm
Weight	1000 Kg

Source

China Coal Mining   Telephone 464847, 464010
  Machinery Manufacture   Cable   6671
  Corp.
21 He Pingli Bei Street
Beijing
People's Republic of China

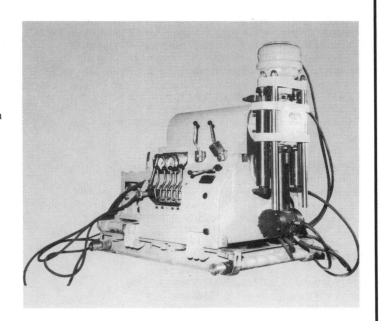

ROCK DRILL

This multi-purpose compressed air drill is suitable for underground drilling during development or production.

  Owing to the weight of these drills they are provided with a jackleg or airleg support, which takes the weight of the drill and provides some thrust. This device is particularly noisy, and ear protection is essential.

Operating information

Drill
— Impact frequency	2600 bpm
— Air supply	141m³ per minute at 6 bar
— Weight	25 Kg without supporting leg

Leg
— Length	1.76m when retracted
— Stroke	1.30m
— Weight	14 Kg

Source

Ingersoll-Rand Company
Rock Drill Division
Phillipsburg
NJ 08865
USA

Other sources

CompAir Holman Ltd   Telephone 06284 6044
Globe Park   Telex   847407
Marlow
Buckinghamshire
SL7 1YB
UK

Atlas Copco AB   Telephone 08 743 8000
Group Headquarters   Telex   14080 copco s
S-10523 Stockholm
Sweden

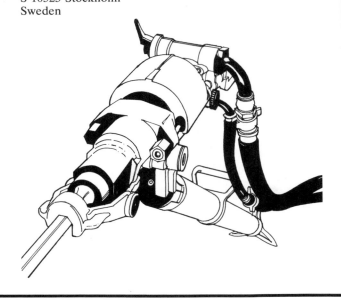

ELECTRIC MOBILE DRILL RIG

Mounted on wheels for mobility, this is a rig for vertical or horizontal drilling underground. The narrow body makes it particularly good for keeping haulages and stopes entirely in-vein. The rig can be used in veins only 1.2m wide and up to sections 3m square. The boom is hydraulically controlled. Electric power is fed via an automatic cable reel system which has a capacity of 85m.

The carrier is a standard base compatible with other equipment from the same manufacturer. The drifter is a rotary percussion machine with pneumatic feed, and is of the high torque type.

Operating information
Carrier:
— Electric motor	50/60Hz, 3 phase, 380/440/ 550V
— Power	22kW at 1450 rpm
— Speed	0 to 7.2 Km per hour
— Maximum gradient	30 per cent

Drill:
— Boom	Hydraulic
— Pneumatic feed length	3200mm
— Stroke	2165mm
— Thrust	258 Kg at 6 bars

Drifter:
— Air pressure	5 to 6 bars
— Weight	50 Kg
— Air consumption	5500 litres per minute

Source
France Loader	Telephone	(1)45 01 81 24
50 Avenue Victor Hugo	Telex	610713
75116 Paris		
France		

Other sources
Mobile drill rigs (development)
Equipment Minier
50 Avenue Victor Hugo
75116 Paris
France

Tamrock	Telephone	358 31 431411
SF-33310	Telex	22193 rock sf
Tampere		
Finland		

Atlas Copco AB
(as previously listed)

Mobile drill rigs (stoping)
Tamrock
(as listed above)

Atlas Copco AB
(as previously listed)

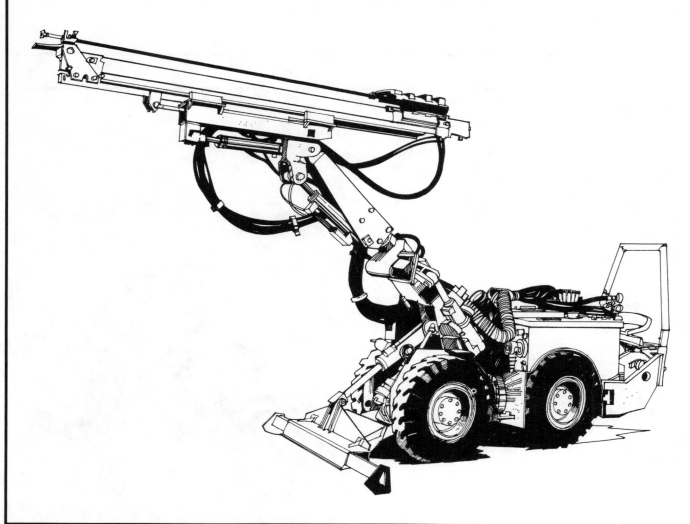

MOBILE UNDERGROUND DRILL RIG

This is a wheeled rig which is designed to increase the mechanization of drilling underground. Its mobility allows it to be moved from one face/heading to another, thus increasing equipment utilization. This particular model requires an opening of at least 3.5m × 3.5m in which to operate.

The rig can drill parallel and single long rings or fans. This makes it suitable for virtually all mining methods. The rig is pneumatic, with a diesel drive available as an alternative.

A vast range of these rigs exists, varying in size, flexibility of drilling pattern, number of booms, etc.

Operating information

Power	Pneumatic or Deutz 57 HP diesel engine
Travel speed	2km per hour (level) 1.5km per hour (8 degree gradient)
Gradeability	20 degrees
Turning radius	3.5m about rig centre point
Feed rotation	360 degrees manually
Feed tilt	20 degrees forwards 20 degrees backwards
Lateral movement	1.5m (for parallel holes)
Length	4.0m
Height	2.4m
Width	2.2m
Weight	6000 Kg

Source

Tamrock
SF-33310 Tampere
Finland

Telephone 358 31 431411
Telex 22193

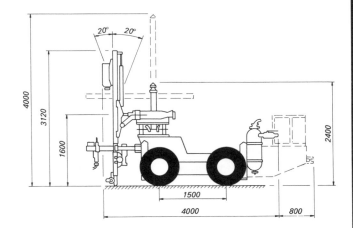

HANDSET HYDRAULIC PROP

A simple but rugged 25-tonne hydraulic prop which gives efficient roof support with low maintenance requirements. Operation is extremely easy and quick.

Such props are used widely to provide temporary support. They result in better roof control and safer underground working.

To raise the prop, fluid is drawn in via a valve from the reservoir. A hand key is used to pump the fluid.

To lower the prop, the same hand key is used, pulling down on the valve and returning fluid to the main reservoir.

Operating information

Load carrying capacity	25 tonnes
Setting load	8 tonnes
Operating height	500 to 2500mm

Source

Guillick Dobson Ltd
PO Box 12
Wigan
UK

Telephone 0942 41991
Telex 67513

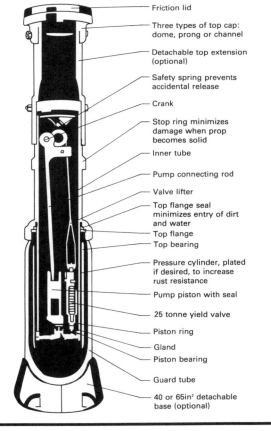

HYDRAULIC CHOCK

The chock is used worldwide, principally as a roof support in underground mines. The chock can also be used as an anchor for securing plant temporarily (for example, the drive head of a chain conveyor). This double acting hydraulic chock is robust and reliable. Valve discs are replaceable. Extension pieces can be used to increase the height, and a number of alternative top and base plates are available to suit roof and floor conditions. These are expensive items of equipment and therefore likely to be useful in larger, more mechanized mining operations.

Operating information
These chocks come in a wide variety of sizes. The following is an example:

Closed height	640mm
Extended height	890mm
Setting load	25 tonnes at 70 Kg per cm^2
Yield pressure	140 to 210 Kg per cm^2
Retract pressure	8 tonnes at 70 Kg per cm^2

Source
Dowty Mining
Equipment Ltd
Aschurch
Tewkesbury
Gloucestershire
GL20 8JR
UK

Telephone 0684 292441
Telex   43285

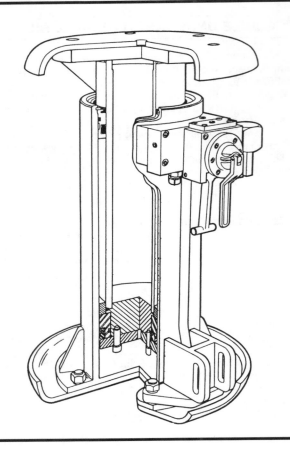

CHOCK RELEASE

Many underground operations use a wooden crib construction to build chocks for roof support. However, once the support has been loaded, controlled collapse is difficult and often dangerous. By incorporating a chock release into the support before it becomes loaded, controlled collapse can be obtained. Also by hitting the steel wedges, the support can be set up to the roof to ensure early support.

Once the chock use has finished or it requires collapsing, the wedges can be released from a safe distance. The chock release can be recovered for re-use. The method also allows more of the timber to be recovered.

Operating information
Chock releases are available in a range of lengths to suit the wooden supports being used.

Essential information on the safe use of chocks and wedges is available in an illustrated leaflet from the manufacturer.

Source
Dowty Meco Ltd
Meco Works
Worcester
WR2 5EG
UK

Telephone 0905 422291
Telex   338370

SLIDE BARS

A slide bar is a box section support bar for use in mines which, because of its section, has high resistance to crushing and lateral bending combined with relatively light weight. Although these bars are traditionally used in stratified deposits, many applications are possible in other mines.

A bar with a serrated edge is available, which allows a positive location into prong-topped hydraulic props.

When used in conjunction with a slide bar bracket, the support can be advanced without having to lower the bar.

Operating information

A number of variants are possible, each with different load characteristics. The details given here apply only to one type, quoted by way of example.

Section depth	86 to 88mm
Section width	100 to 101mm
Weight	26 Kg per metre length
Crushing resistance	1100 KN minimum
Brinell hardness	197 to 255
Section modulus	
— XX axis	76.5m³
— YY axis	49.5m³

Source

Becorit Ltd	Telephone	0602 302603
Hallam Fields Road	Telex	37526
Ilkeston		
Derbyshire		
DE7 4BS		
UK		

ROCK STABILIZER

This stabilizer is a simple but effective alternative to roof bolts, with their high maintenance time (for example, with regard to torque adjustment).

There are two parts to the stabilizer, a tube and a roof plate. The tube has a longitudinal slot, so that the tube is compressed when it is driven into an undersized hole. Conventional roof bolts tend to work loose with rock movement but this stabilizer deforms with the rock and maintains support.

As no additional maintenance is needed, considerable cost savings can be achieved.

Operating information

Diameter	33mm, 39mm and 46mm versions
Lengths	To suit the mine's requirements

Source

Ingersoll-Rand Co.	Telephone 609 921 8688
Split Set Division	
100 Thanet Circle	
Suite 300	
Princeton	
NJ 08540-3662	
USA	

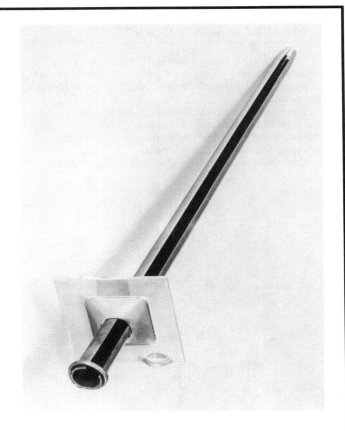

CHEMICAL CARTRIDGE FOR ANCHORING, SEALING AND DOWELLING

These cartridges can be used for the partial or full sealing of anchoring rods in boreholes (such as roof bolting), dowelling of support elements, securing coalfaces and rock formations. They are also suitable for general fixing applications, such as dowelling heavy suspension gear, or for machinery foundation bolts.

The resin is thixotropic (which means that it will not flow out of a roof hole) but it will deform to fill the space between the bolt and the hole wall. Cartridges can be inserted pneumatically, and the resin hardens fairly rapidly to secure the dowel or bolt in the hole.

This product has a shelf life limitation of six months (minimum) at 25 degrees Celsius.

Operating information
There are two standard diameters, of 23mm and 28mm. Each of these is available in lengths of 300, 500, 600 and 750mm.

Cartridges are supplied in standard pack quantities of 30 and 20 for the 23mm and 28mm items, respectively.

Standard setting time at 25 degrees Celsius can be specified according to the following type of cartridge:

Type 0.6 SF	20 to 28 seconds
0.8 SF	40 to 45 seconds
3 SF	3 to 4 minutes

Other lengths, other setting times, and other diameters between 24 and 27mm can be specified.

Source
Consulenza Appalti Edili   Telephone 02 6466678
 Minerali (CAEM)   Telex   310579
Via Rapisadi 9
20162 Milan
Italy

BALANCED SHAFT DOORS

A single weight, ropes and pulleys are used in a counterbalancing arrangement for opening and closing these double doors, which are designed for the top of a vertical shaft.

The doors are mounted with the hinges placed so that their pin centres are above rail level. This allows the rails to meet with little gap when the doors are closed, ensuring a smooth ride for wagons.

The whole arrangement is set into the shaft collar.

Operating information
Dimensions and the strengths of materials used depend on the size of the shaft collar.

Source
Fabricate locally.

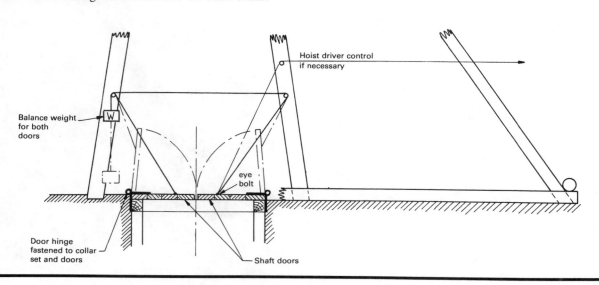

Balance weight for both doors — W

Hoist driver control if necessary

eye bolt

Door hinge fastened to collar set and doors

Shaft doors

MINE VENTILATION

The fans illustrated here are part of a vast range available from this manufacturer, who supplies centrifugal and axial fans for surface buildings and underground installations worldwide. For any ventilation requirement the reader is advised to consult the manufacturer, who has establishments or is otherwise represented in 70 countries throughout the world.

The aerofoil jet range of fans are self contained units producing high thrust from relatively low electrical input. They are constructed for bolting to tunnel roofs, and are available in a range of sizes, among which are 630, 1000 and 1250mm diameter units. Corrosion protection is normally by galvanizing, but epoxy paint finish is also available if needed.

A standard range of axial aerofoil fans is produced in sizes from as small as 300mm up to nearly 3 metres diameter. These are galvanized for protection against corrosion, and can be used in combinations. A full range of matching ancillary equipment is available.

Known as the Varofoil range, the company makes axial fan assemblies which have variable pitch-in-motion blades. These conserve energy since the airflow can be regulated almost to zero flow when appropriate. Energy reductions are 50 per cent for an air flow reduced to 80 per cent of maximum rating and, with the air flow reduced to half, the electrical consumption drops to only 13 per cent of full load. There are 13 sizes of fans in this range, from 630mm diameter right up to 2800mm. The largest unit can supply air at 150m³ per second, and the highest static pressure rating is 2kPa. As with other types of fans, higher volumes and pressures can be obtained using multi-staging and parallel arrangements.

Operating information

Owing to the wide range of possibilities, it is necessary to consult the manufacturer for specific requirements.

It should be noted that this company has extensive experience of ventilating areas contaminated by dust, ash, grit and other abrasive materials.

Source

Woods of Colchester Ltd
Tufnell Way
Colchester
CO4 5AR
UK
This company is part of the GEC group.

Aerofoil jet fan

Standard aerofoil unit

A varofoil fan unit

MINING SUSPENSION THEODOLITE

A mining theodolite is essential for surveying angles in underground workings. The attractive feature about this suspension theodolite is that it does not need a tripod, because it can be suspended from the roof. The design makes it useful in steep or narrow openings.

The point of suspension may be a steel punch set into a wood support. A ball suspension joint and two levelling screws used in conjunction with the single-bubble circular spirit level allow the theodolite to be set up quickly.

Special supports are available that allow this instrument to be mounted on a tripod, extending its use to many surface surveying applications.

The price is variable, according to accessories required.

Operating information

Optical magnification	18×
Objective aperture	30mm
Graduation	360° in one degree intervals
Scale interval	5 minutes
Reading definition	30 seconds (by estimation)
Weight	3.2 Kg without case

Note that many accessories are available for either essential or optional use with this instrument, and the manufacturer should be consulted for details.

Source

F.W. Breithaupt &   Telephone 0561-772042
  Sohn GmbH   Telex   992258
PO Box 100569
D-3500 Kassel
West Germany

Other sources

Wild Heerbrugg Ltd	Telephone	71 70 3131
CH-9435 Heerbrugg	Telex	881221
Switzerland		
Kern & Co. Ltd	Telephone	64 26 4444
CH-5001 Aarau	Telex	981106
Switzerland		
Carl Zeiss	Telephone	7364 203242
D-7082 Oberkochen	Telex	713751
West Germany		

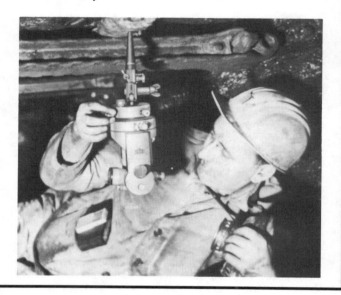

MINE COMMUNICATIONS SYSTEM

The communications system described here is designed to intrinsically safe standards for use underground in mines, including those which are gassy.

The system comprises loudspeaker amplifiers which can be linked by cable. Individual units can be used for two-way speech and to call an exchange. Alarm tones can be received, and linked to the start-up of certain items of plant.

A power supply unit is required, and one of these is capable of supplying up to 25 loudspeaker amplifiers. Battery back-up power is provided for each unit, allowing between 6 and 130 hours' operation, depending on battery condition. The power unit comes with a power supply monitor, battery condition monitor and amplifier.

Operating information

Each loudspeaker amplifier:

— Supply	12V dc. from the power unit
— Back-up	8.4V at 600mAh, provided by six rechargeable nickel cadmium cells which are charged from the 12V power unit supply
— Audio power	2 W, giving a sound level of 80dB measured at 1m distance from the loudspeaker
— Cable	4-core, 2.5mm square SWA communications cable (not supplied)

Source

NEI Mining   Telephone 0283 43471
  Equipment Ltd   Telex   34654
Shobnall Street
Burton on Trent
Staffordshire
DE14 2HD
UK

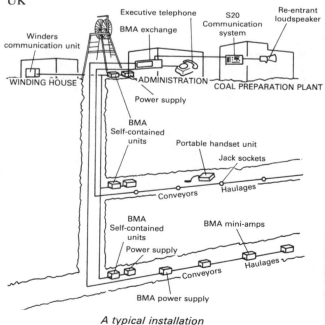

A typical installation

4. Mineral processing

The equipment contained in this section may be utilized for processing a variety of ores. The equipment is suitable for processing run-of-mine ore at approximately 150mm into products, typically concentrates for further treatment, and a waste material for disposal. The equipment falls into the following categories: crushers, mills, classifiers, gravity separation devices, flotation, filtration, mobile units, and laboratory equipment.

The type of equipment to be selected from each category will be solely dependent upon the nature of the ore, and the recovery method to be used for the value mineral. It is essential therefore that the method of recovery is developed with approximate tonnages prior to attempting to select the individual process equipment.

Each process section is discussed below as are the parameters which must be considered prior to equipment selection.

Crushing

The run-of-mine (ROM) ore must be reduced in size in order to be suitable as feed to the downstream process. The product sizing is usually of the order of 10 to 20mm and may be achieved by multi-stage crushing, as follows:

Primary crushers

These are usually gyratory or jaw crushers. Jaw crushers are lower cost per unit throughput but are unsuitable for slabby feed material. Typical ratio of reduction for both crushers would be four to one.

Secondary and tertiary crushers

These are usually gyratory or cone type, or if the material is exceptionally cubic a secondary jaw crusher could be used. The final stage of the crushing should be in closed circuit with some form of screen to recirculate oversize product back to the crusher. Again, ratio of reduction is usually three or four to one.

Single stage crushers

(i) As an alternative to a two stage crushing system a single stage (rotary) impact crusher can be used. Power consumption and wear costs are usually high on these items unless the ore is soft. They do however handle sticky ores very well.

(ii) A stamp mill may be used if a suitable unit can be obtained from some second-hand source. They are no longer produced and can only be obtained from redundant operations. They are very low throughput units and relatively inefficient.

Roll crushers

These are used to prevent the production of fines while generating an accurate top sized product. Toothed rolls are generally primary crushing units while smooth rolls are secondary or tertiary units. Suitable for cubic feed and multi-stage crushing of friable ores.

Grinding mills

These generally fall into three categories for reduction of ore to nominally 0.1mm.

Autogenous mills

These are large diameter short length mills which can be used for coarse feed which breaks on itself during tumbling. The ore must have the value mineral associated with grain boundaries or fracture planes, and the host rock must be amenable to autogenous grinding. Autogenous mills are usually only suitable for feed rates above about 10 tonnes per hour.

Rod mills

With these, length must be 30 per cent larger than the diameter to prevent rod tangling. They can take a coarser feed than ball mills (20mm) and produce a coarser product with less fines. They are usually operated in open circuit when feed is clayey and would cause problems in fine crushing, as preparation for secondary ball milling.

Ball mills

To use these, the L:D ratio is relatively unimportant. The feed size must not exceed 10mm and the product is usually classified with the oversize returning to the mill (i.e. in closed circuit). Final product size is generally finer than rod mills and can be as low as 20 microns when used as a regrind mill.

Pebble mills

The length of these mills is usually larger than the diameter. Ore is fed as fines plus pebbles and the pebbles are used as the grinding media. These mills are preferable where iron is a contaminant in the mineral product. They achieve a relatively low throughput per unit volume compared with rod/ball mills.

Classification

This may be considered in two forms: (a) coarse rock classification on shaking screens and grizzleys down to approximately 0.5mm; (b) fine ore classification usually as a slurry in mechanical or centrifugal classifiers down to one micron.

Coarse ore

(i) *Grizzleys* are usually installed ahead of a crushing unit to allow undersize to bypass the crusher — normally when at least 25 to 30 per cent of the feed can be bypassed around the crusher.

(ii) *Shaking screens* are used to limit the size of material passing to the following process. Screen decks can be made of a variety of material e.g. steel, rubber, polypropylene, etc., to allow a trade-off of capital and operating cost. Screen openings may be square, rectangular or slotted depending upon the ore characteristics.

Fine ore

Cyclones depend upon centrifugal force to cause a separation of particles of differing mass. They can be used to separate from nominal 1-5 microns up to 300-500 microns. In general, the larger the cyclone the coarser is the cut point. Operation is dependent upon a steady feed, usually pumped.

Mechanical classifiers. Rake or screw classifiers can accept a surging feed and usually a coarser feed than a cyclone. They are more expensive than a cyclone but experience fewer operating problems and do not need a feed pump.

Gravity separation

Separation is dependent upon the difference in specific gravity causing a partition to be formed. It is evident therefore that to use these processes requires the value mineral to be free and of sufficient difference in density from the host rock. These units will then separate particles by mass thus generating both free mineral and waste rock of, say, half the density but twice the size. Subsequent sizing and re-concentration is used to generate high grade concentrate.

Minimum SG difference that can be separated is usually 0.5 but ideally should be 1.0. The units are generally feed size specific and may be classified accordingly:

Wet separators

(i)   Jigs from 20mm down to 0.5mm.
(ii) Spirals from 2mm down to 40 microns.
(iii) Tables from 1mm down to 40 microns.
(iv) Slimes tables from 100 microns to 10 microns.
(v)   Pinched sluices, Chinese launders, (Palangs) sluice boxes, etc. being a crude form of tabling.
(vi) Heavy media from 25mm down to 0.5mm

Dry separators

These are used specifically where water would detrimentally affect the product, or where water is unavailable. Basic principles are identical to wet processes but use air as a fluidizing medium. Again the separation is by mass and as such the feed should be carefully sized, unless the value mineral is in a small range of particle sizes.

Flotation

This process depends upon the difference in surface characteristics between the ore and the waste when treated with reagents. One is made to be attracted to air while the other is attracted to water. The separation is made by collecting the product in an aerated tank and collecting the float product into a stabilized froth which overflows the cell. Flotation is usually carried out on material below 500 microns, although its effectiveness drops off below about 25 microns and is usually ineffective below about 10 microns.

If the ore is amenable to flotation then the process is usually more expensive but more efficient than gravity concentration.

Filtration

The products from gravity concentration or flotation are inevitably in the form of a slurry which requires dewatering.

This is carried out in settling tanks (thickeners) following which the thickened slurry is usually filtered to approximately 10 per cent moisture.

Small-scale filters usually take the form of tipping pan filters operated on a batch basis. Continuous and semi-continuous filters take the form of drums, discs, horizontal pan, horizontal belt, plate and frame, and pressure filters.

The following generalizations apply to the selection of each type.
(a) Drum filters are suitable for free draining concentrates capable of slurry suspension in the filter bath. They are not suitable for coarse-grained industrial minerals.
(b) Disc filter applications are similar to drum filters but provide larger filter area per unit of floor occupied.
(c) Horizontal pan, and belt. Suitable for coarse products which would not form a coherent stable cake on a vertical filter. Capital costs are higher and these filters require more floor space.
(d) Plate and frame are used for fine sticky concentrates which require pressure as opposed to a vacuum in order to form a cake.
(e) Pressure filters are used where additional pressure is required to achieve the desired moisture level, usually for long haul transportation or reduced drying costs.

Mobile units

Within reason, any combination of the preceding equipment can be trailer mounted into a mobile unit. This can take the form of an extremely small integrated plant, or larger plants can be broken down into unit processes and installed on numerous trailers for reassembly at site using flexible piping. The limitations are usually trailer size, axle load and site access.

Laboratory equipment

Specific small-scale mineral processing equipment is available to duplicate process and pilot plant equipment at bench scale level to test the selected process. It should be emphasized that this equipment is designed for batch laboratory work and is *not* continuously rated small-scale process equipment.

Len Holland

STAMP MILL

Although obsolete in many parts of the world, stamp mills are effective for small-scale operations run by 'artisanal' workers.

Stamp mills are simple and robust machines which accept a coarse feed, and their operation is readily understood. They are essentially suited to the production of granular material from ores containing coarse mineralization.

Stamp batteries are designated by the number of stamps and their falling weight (still using imperial measures). A 3 × 750 mill will therefore have three stamps, each weighing 750 lbs.

Operating information

While a stamp mill can handle material of up to 100mm size, 50mm will be the usual maximum for the lighter units used in small-scale mines.

The position of the tappet on the stem (see diagram)

can be adjusted to give a high drop for hard ores and a shorter travel for soft ores, thus minimizing wear on the shoes and dies.

The discharge screen is the most positive control of capacity: the coarser the screen, the greater the output. The screen is measured in 'battery mesh', which is the number of holes per square inch. A battery mesh of 600 is preferred (equivalent to approximately 0.7mm hole size).

Source

New units are seldom made, as existing ones seem to go on for ever, and the most likely source will be to find a second-hand unit. The following consulting firm has considerable experience in these machines, and more information can be obtained from them.

John Holloway
and Associates
PO Box 5438
Harare
Zimbabwe

Telephone Harare 706174
Telex   2195
(answerback JHAPBS ZW)

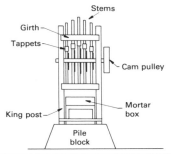
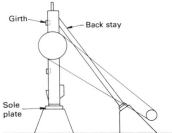

SINGLE DRUM HORIZONTAL SIZER

The operation of this sizer is similar to that of a twin drum machine, but with reduced size and power requirements. It is suitable for materials up to 90 MPa.

The drum is fitted with replaceable picks. A breaker plate intermeshes with the drum teeth to create the second breakage surface, eliminating the need for a second drum. The gap is adjustable, allowing output sizes to be varied.

Operating information

Maximum capacity   120T per hour
Maximum feed size   230mm
Minimum product size   minus 50mm
The unit is supplied complete with a 75 HP TEFC motor, fluid coupling and gear reduction unit.

Source

BQ Mining Industries Ltd Telephone 021 421 606
Mining Equipment   Telex   335198
  Division
Adams Hill
Bartley Green
Birmingham
B32 3EB
UK

All manufacturers listed here manufacture primary, secondary and tertiary crushers.

Other sources

Allis Chalmers
PO Box 512
Milwaukee
WI 53201
USA

Symons Corp.
1155 Churchill Drive
New Braunfels
TX 78130
USA

Pegson Ltd
Coalville
Leicester
LE6 3ES
UK

KHD Humbolt Wedog A.Co.
Postf. 910457
D-5000 Köln 91
West Germany

Roxon
Keskikankaantie 9
SF-15860 Hollola
Finland

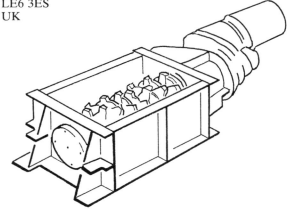

GRINDING MILL

Useful for pilot plants or for small-scale production, this mill is of the continuous feed and discharge type.

The mill drum rests on four pneumatic-tyred vehicle wheels, eliminating the need for special foundations and trunnion bearings. One pair of these wheels is motor driven through a flexible coupling and gearbox. Rollers bear on the inner surfaces of the drum's end flanges to restrain axial movement.

Rod, ball or pebble mill options are available, with a choice of overflow, grate or peripheral discharge arrangements. Scoop, drum, spout or combination feeders are available for wet grinding and chute or vibratory feeders for dry milling.

Operating information

Mill diameter	914mm
Length	1524mm
Lining	Rubber with lifter bars (steel or ceramic are other options)
Power	Typically 15 HP at 725 rpm
Typical throughput	Nominal 2 tonnes per hour

Source

GEC Mechanical
 Handling Ltd
Cambridge Road
Whetstone
Leicester
LE8 3LH
UK

Telephone 0533 863434
Telex   347344

This mill is manufactured as a range that can be adapted to widely varying duties. For prices and further information the manufacturer should be consulted.

Other sources

Allis Chalmers
PO Box 512
Milwaukee
WI 53201
USA

Loesche GmbH
Steinstrasse 18
D-4000 Düsseldorf
West Germany

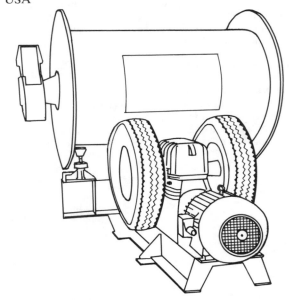

CONTINUOUS ROD/BALL MILL

A rod/ball mill with continuous feed is suitable for primary grinding after a crusher.

The mill rotates while the charge (steel balls/rods or pebbles) grinds the feed. A much finer product is obtainable when compared to a crusher.

Grinding is often necessary to liberate a valuable mineral (which may be finely disseminated) from the gangue.

Although grinding is usually carried out wet (reducing energy) this mill can also be used dry. It comes complete with drum feeder and overflow trommel screen.

Operating information

Power	5 HP
Ball/rod charge	340 Kg
Capacity	136 Kg per hour
Diameter	406mm
Length	1220mm

The above performance is based on wet grinding a medium hard ore from ¼ inch-65 mesh. Dry grinding capacity is about 70 per cent of that for wet grinding.

The internal linings and lifter bars (wear items) are easily replaceable. Mill circuits vary considerably in order to obtain the desired product.

Source

Sepor Inc.
718N Fries Avenue
PO Box 578
Wilmington
CA 90748
USA

Telephone 213 830 6601
Telex   194456

Other sources

Allis Chalmers
PO Box 512
Milwaukee
WI 53201
USA

Pegson Ltd
Coalville
Leicester
LE6 3ES
UK

GEC
Rugby
Warwickshire
CV21 1BD
UK

Fives-Cail Babcock
7 Rue Montalivet
F-75383 Paris Cedex 08
France

CANTILEVER GRIZZLEY

A grizzley is a coarse screen which is used principally to:
1. Separate fines from a crusher feed (the grizzley should be sloped)
2. To hold back lumps that are too large to enter the crusher.

Simple grizzleys can be made from bars at fixed spacing, but moving bar types are far more efficient. The cantilever grizzley is one way of mechanizing small plants.

The grizzley bars are made from old 20 lb rails with their flanges removed (with the bulbs facing upwards). The web is drilled approximately one-third of the total grizzley length to take a fulcrum rod. The fulcrum rod allows the rails to be mounted a fixed distance apart. For strength, extra pieces are welded to the web. Each rail is free to move independently, but each is limited by the bar guide.

Operating information
Being pivoted close to the feed, the bars are overhung and free to vibrate at the discharge end. Thus ore loading causes each bar to move separately at the opposite end, reducing blockages and encouraging the oversize to move forwards.

Source
Usually designed and fabricated to suit individual requirements.

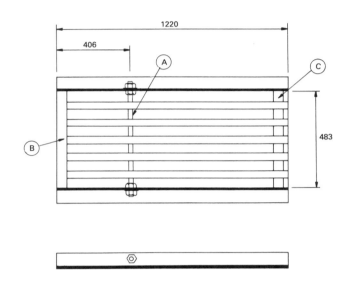

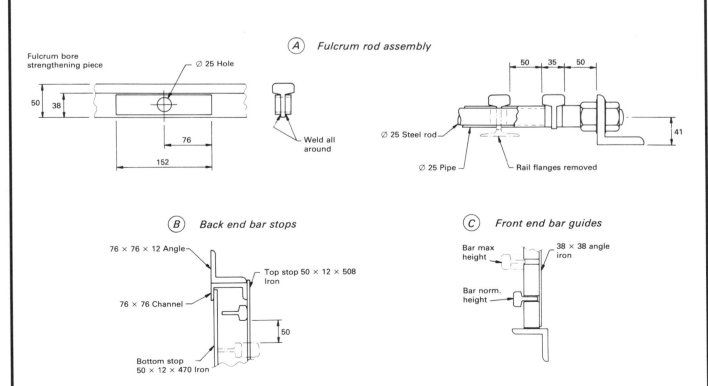

(A) Fulcrum rod assembly

Fulcrum bore strengthening piece
Ø 25 Hole
50 38
76
152
Weld all around

50 35 50
Ø 25 Steel rod
Ø 25 Pipe
Rail flanges removed
41

(B) Back end bar stops

76 × 76 × 12 Angle
Top stop 50 × 12 × 508 Iron
76 × 76 Channel
50
Bottom stop 50 × 12 × 470 Iron

(C) Front end bar guides

Bar max height
38 × 38 angle iron
Bar norm. height

HORIZONTAL VIBRATING SCREEN

This machine is designed with low height accommodation requirements and can be supplied with a range of screen sizes and combinations.

Screens can be supplied in wedge wire, woven wire, perforated plate, loose rod, resilient rod deck, etc. construction. The decks are made in a variety of abrasion-resistant materials, such as stainless steel, polyurethane, high carbon steel, and so on.

Each screen is excited by a factory-assembled vibrating mechanism, which is mounted on the machine in a manner that stiffens the side plates. Two eccentric steel shafts are geared together, producing a straight line motion at 45 degrees to the screen surface which passes through the machine's centre of gravity. The frequency and amplitudes of vibration are chosen to suit the application.

High efficiency is claimed for the separation process. Particles are thrown in a forward and upward direction, which ensures multiple exposures to the screening surface.

Operating information

Screen sizes	From 0.9 × 2.4 metres up to 3.0 × 7.2 metres
Decks	Single, double or triple options in all sizes
Drive	Electric motor mounted on or close to the vibrator mechanism, driving through a belt
Throughput	Solely dependent on size distribution of feed and deck aperture

Source

Magco Ltd
Jenkins of Retford Ltd
Thrumpton Lane
Retford
Nottinghamshire
DN22 7AN
UK

Telephone 0777 706777
Telex   56122

PORTABLE VIBRATING SCREEN

This electrically-driven vibrating screen is suitable for batch or continuous screening of wet or dry material from 1 inch to 200 mesh.

The screens can be changed easily and arranged as single or double deck. Screens can be mild or stainless steel, as required.

A stand, fitted with castors, can be purchased. This allows the pitch to be adjusted from 10 to 40 degrees.

The feed material is delivered from a chute which is controlled by a gate. The output is via another chute into a pan.

The deck can be open or closed. The closed version is dust tight, and should be specified whenever environmental controls are imposed, or if the dust could be dangerous to personnel.

Operating information

Motor	110 volt or 220 volt, single phase, 60 Hz. Others are available.
Size	24 × 36 inches (60 × 90 cm)
Throughput	Dependent upon deck aperture and feed size distribution.

Source

Sepor Inc.
PO Box 578
Wilmington
CA 90748
USA

Telephone 213 830 6601
Telex   194496

Other sources

Envirotech Corp. (Wemco Division)
PO Box 15610
Sacramento
CA 95813
USA

Denver Equipment Division
 (Joy Process Equipment Ltd)
9 North Street
Leatherhead
Surrey
KT22 7AY
UK

Allis Chalmers, Pegson and GEC (all addresses given earlier) also manufacture vibrating screens.

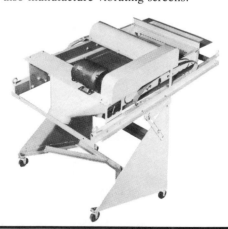

HYDROCYCLONE

Hydrocyclones are simple, but extremely effective mineral processing systems. Slurry feed is introduced to the cyclone tangentially. The resultant motion and forces within the unit create the conditions necessary for separation.

The separated products report either to the vortex finder at the top or to the spigot at the bottom. Hydrocyclones are efficient and can give sharp cut points. They are manufactured from abrasion resistant polyurethane, and the designs are flexible to allow a variety of geometrical set ups.

Installations vary from the very simple to the complex. The amount of additional equipment required (such as pumps and piping) depends on the circuit.

Operating information
Applications include:
— Classification of solids within a slurry feed
— Beneficiation of industrial minerals
— Separation of solids from liquids
— Thickening of slurry feed
— Degritting of solids
— Desliming of solids

Source
Richard Mozley Ltd
Cardrew
Redruth
Cornwall
TR15 1SS
UK

Telephone 0209 211081
Telex   45735

Allis Chalmers, Fives-Cail Babcock, KHD Humbolt Wedog, Denver Equipment Division (all addresses listed earlier) also supply cyclones.

Prices vary considerably, depending on the size of the cyclone and the amount of additional equipment required.

HAND SCREEN

Screens are used before many processes to size the feed closely, resulting in increased recoveries. The simple screen shown here can be made locally.

The screen is suspended on a pivoted sling, which allows it to be hand vibrated.

Such devices are only suitable for sizing down to 4mm.

Operating information
This version is hand operated, but it can be mechanized by coupling it to a suitable engine with an eccentric or cam drive.

Screening can be undertaken wet or dry. Wet screening reduces blinding, but then a water supply obviously needs to be provided.

Source
Manufacture locally to suit requirements.

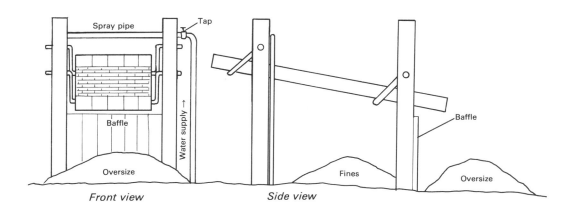

SELECTIVE MINERAL JIG

This single-compartment hutch mineral jig can be used in continuous or batch mode. It is suitable for pilot plant or laboratory operations and in placer clean up. Usually the unit is used for sands, from coarse to very fine.

The jig is a mechanical concentrator which works on a gravity principle and needs a constant supply of water. Heavy grains are separated from the light particles by their ability to penetrate a semi-stationary bed acting through a pulsating fluid.

Stainless steel is used for the screen, neoprene for the diaphragm, and the frame and stand are fabricated from steel.

Operating information

Capacity	1 tonne per hour
Feed size	⅜ inch (about 9mm)
Diaphragm stroke	Adjustable between $^3/_{32}$ inch and $^{13}/_{16}$ inch (approximately 0.2mm to 13.3mm)
Power	Electric motor, ½ HP, 115 or 230 volts, single phase, 60 Hz

Uses include separation of black sands, precious minerals, lead, zinc and for the cleaning of non metals and coal. The desired recoveries and grades are obtained by altering five operating parameters:
— Speed
— Stroke
— Screens
— Bed material
— Water dilution.

Source

Sepor Inc.   Telephone 213 830 6601
718N Fries Avenue
PO Box 578   Telex   194496
Wilmington
CA 90748
USA

Other sources

Mineral Deposits Ltd
Box 5044, Gold Coast Mail Centre
Bundall 4217, Queensland
Australia

Wilfley Mining Machinery
Cambridge Street
Wellingborough
Northamptonshire
NN8 1DW
UK

Mozley Ltd
Cardrew
Redruth
Cornwall
TR15 1SS
UK

GEC and Envirotech Corp (Wemco Division) also supply 'wet' gravity separation equipment. Both addresses are listed in previous entries.

HYDRAULIC RADIAL JIG

This hydraulic drive jig is suitable for use on mining dredges. It can be used for primary separation and first stage cleaning. The primary use of this unit is for processing placer gold and tin ores. The table is driven in a pulsating motion which has a saw-tooth waveform.

This manufacturer also produces a wide range of other concentrating and mineral treatment equipment, including small jigs and sluices.

(NB: This jig is suitable for larger-scale mining operations only.)

Operating information

Drive	Mechanical-hydraulic
Outside diameter	7.75m
Number of compartments	6, 9 or 12
Jig bed area	3.3m² per compartment
Feed size	Minus 25mm
Capacity	100 to 150 tonnes per hour (six compartments version)
Hutch water	About 1.2 tonnes per tonne of ore
Water pressure	50 KPa
Mechanical drive power	7.5 kW motor
Hydraulic pump power	0.6 kW motor
Weight	10 tonnes with six compartments 13 tonnes with nine compartments

Source

Beijing General Research   Telephone 890531
   Institute of Mining &   Telex   222589
   Metallurgy (BGRIMM)
1 Wenxing Street
Xizhimenwai
Beijing
People's Republic of China

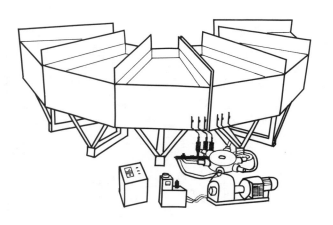

WATER JIG

The water jig is a lightweight machine which is used for fast and accurate analysis of the heavy mineral content of samples. Such samples may be from beach sands, placers and alluvial deposits. It can also be used for the separation of heavy minerals (for example, diamonds, gold and cassiterite).

The jig is motorized for a faster analysis cycle time, but a cheaper manual model is available. Although slower than its motorized counterpart the manual version does enable a systematic search to be made, and is less tiring, quicker and more reliable than the traditional wash trough.

Operating information

A filtering bed with glass balls provides separation of light gangue (3.0 SG) and heavy minerals when the density difference in relation to the gangue is small (diamonds, for example). The motion can be controlled to suit the feed characteristics.

Source

BRGM Instruments Telephone 38 64 34 18
BP 6009 Telex 780258
45060 Orleans Cedex 2
France

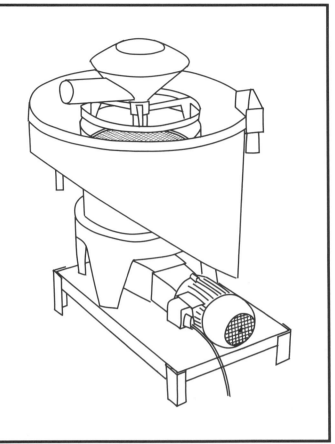

SPIRAL CONCENTRATOR

The spiral concentrator is a low capital, low operating cost unit which employs gravity, centrifugal and drag forces to separate minerals of differing specific gravities in a flowing pulp stream.

In operation, the heavier minerals settle and are retarded in the lower stream layers. These are progressively removed in concentrate collection ports located at the inner edge of the channel.

Minerals treated include: apatite, barite, cassiterite, chromite, coal, columbite, diamond, fluorite, galena, gold, ilmenite, monazite, pyrite, rutile, scheelite, tantalite, uraninite, wolfram and zircon.

Operating information

The makers of this unit claim advantages over similar currently available equipment, as follows:
— Reduced water consumption, with up to 8 gallons per metre (36.5 litres per metre) less per spiral.
— Improved recovery, particularly of fine heavies, due to the reduced wash water requirements, with no loss of grade.
— Higher capacity, with up to 100 per cent increase in throughput. (Throughput nominally one to one-and-a-half tonnes per hour per spiral start.)

Source

GEC Mechanical Telephone 0533 863434
 Handling Ltd Telex 347344
Cambridge Road
Whetstone
Leics
LE8 3LH
UK

A large GEC spiral concentrator installation

GOLD CONCENTRATOR

This is a portable gold concentrating device for testing or sampling placer deposits and for small-scale production.

The principal components of the unit are a revolving scrubber and trommel screen, feed hopper, sprays, pump, concentrating riffles and petrol engine.

The revolving scrubber and trommel screen are fed from a hopper, and oversize material is rejected at the screen. The undersize material is then fed to a vibrating riffle sluice box, where the concentration occurs.

A considerable flow of water is needed, but the closed circuit allows recycling.

Operating information

Power	Petrol engine, 2 HP, air cooled with centrifugal clutch
Capacity	2 to 3 cubic yards per hour (about 1.5 to 2.3 cubic metres per hour)
Dimensions	1.52m × 1.17m high
Weight	268 Kg (largest single piece 80 Kg)

Source

Sepor Inc.
718N Fries Avenue
PO Box 578
Wilmington
CA 90748
USA

Telephone 213 830 6601
Telex   194496

LIGHTWEIGHT GOLD WASHING DEVICE

Dubbed the 'Gold Genie' by its manufacturer, this is a very lightweight gold washing machine which is faster and easier than hand panning. Power is supplied by a small 12 volt motor operating through a belt drive, and the whole unit is supported in a collapsible frame that can be packed into the boot of a car.

An electric pump is supplied with the unit for drawing the washing water from a river or other local source. Water is pumped to the bowl through a spray bar. The bowl is made from polyurethane for lightness and durability. The pale colour of this bowl is claimed to be particularly useful for showing off gold and other colours.

The washer is designed for reclaiming fine gold from a previous process. It can be used directly after the sluice box of a dredge, with the sluice feeding directly to the bowl. In addition to gold, other heavy minerals can be separated, such as silver, cinnabar, mercury, gemstones, platinum, lead, galena, manganese and zinc.

Operating information

Feed material	6.4mm if concentrated 38mm for river run material
Power required	12 volts dc (battery not supplied)
Motor current	2.5 Amperes
Pump current	2.0 Amperes
Weight	25 lbs (11 Kg)

The operating parameters which have to be adjusted to obtain maximum recoveries at the desired grade are:
— bowl angle
— spray rate
— speed of rotation.

Source

The UK distributor of this equipment is:
Christison Scientific
  Equipment Ltd
Albany Road
East Gateshead Industrial Estate
Gateshead
NE8 3AT
UK

Telephone 0632 774261
Telex   537426

SMALL WASHING UNIT

The familiar technology of a concrete mixer and a pump are used in this highly mobile washing unit. Apart from its application as a washer, it can also be used for deposit evaluation by bulk testing. The constituents are:
— Motorized pump
— Concrete mixer
— Wash table, with two water spray bars
— Sluice, with a mat at the bottom.
Adjustable feet are provided for levelling on all types of surface.

Operating information

Capacity	0.5 to 1 m³ per hour of auriferous alluvium and eluvium
Pump capacity	40 m³ per hour
Mixer capacity	150 litres
Sluice length	4 m, with riffles, 0.1 m apart
Fuel consumption	2 litres per hour
Total weight	280 Kg

Source

BRGM Instruments   Telephone 38 64 34 18
BP 6009   Telex   780258
45060 Orleans Cedex 2
France

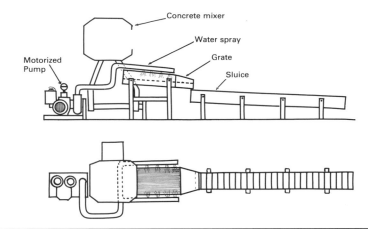

HAND OPERATED CONCENTRATING TABLE

Riffled concentrating tables are particularly useful for cleaning concentrates from other processes. Commercially produced tables, together with the necessary drives, are costly. Tables can, however, be made locally from wood.

Separation results from the combined motion and action of the riffles. A flywheel helps to produce the motion, which consists of a backwards and forwards table oscillation with a kick caused by the end stop. The rope supports the table without restricting the motion.

The riffles on the table are shorter at the feed side, and all taper from left to right on the diagram. Wash water is supplied through a perforated pipe, which serves as a spray bar. Collection launders can be arranged to catch the required products.

Operating information

Dimensions are not given in the illustration because these will depend on factors such as the throughput required. A full-sized unit nominally 2m by 5m will treat approximately 2 tonnes per hour of minus 1mm feed.

Source

Can be fabricated locally to suit particular requirements.

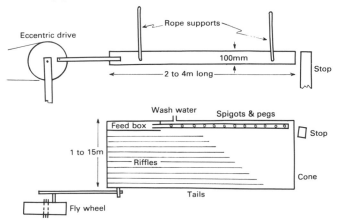

ROCKER BOX

The rocker box is a simple device for the processing of heavy minerals from alluvial deposits, especially gold.

The rocker consists of a trough with riffles behind which the heavy mineral is caught and recovered. The riffles are mounted on the base of a box construction which, in turn, is mounted on curved bars that allow the whole construction to rock.

The riffles are fed from a hopper, fitted with a screen on the underside to guide the material.

Fine gold from the rocker is caught from underneath in a piece of cloth.

Operating information

The device can be operated by one man, but two are more efficient.

Separation of the sand from gold is achieved by shaking the rocker vigorously while a steady stream of water flows over the riffles.

Typical capacity	2m³ in 10 hours, with two men
Water required	600 to 1100 litres for the above throughput

The water can be recycled via a resettling tank.

Source

Illustration is taken from *Queensland Government Mining Journal*, December 1977. Can be fabricated locally to meet particular requirements.

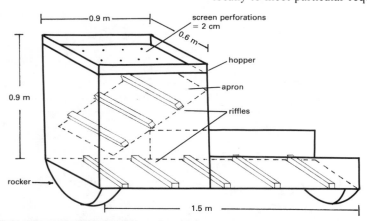

WILLOUGHBY CONCENTRATOR

A Willoughby concentrator can be used to further upgrade alluvial tin or wolfram deposits.

The ore box is 305mm deep. An integral water tank/tower, 1.52m high and 508m in diameter, is joined to the box by a passage which is 457mm square. The base of the ore box has a finely perforated plate.

The whole construction is contained within a box measuring 1.83m high by 762mm square. Valves from this sump allow the introduction and removal of water.

The separation process is similar to that of a batch hand jig.

Operating information

Capacity	2 tonnes of enriched ore per 10 hour shift

The operating procedure is as follows:
1. Charge the ore box with closely sized ore — 20#.
2. Fill the water tower.
3. Open the bottom valve from the water tank, allowing water to pass out and over the ore bed (5 to 10 minutes).
4. Close the tank valve, and open the valve below the ore box, allowing the water in the ore box to drain into the sump. The tower capacity must be sufficient to prevent the water level exceeding the sieve level.

Source

Fabricate locally to suit requirements.

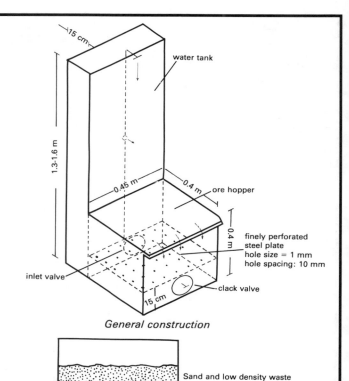

General construction

Material separation in the hopper

STREAMING BOX

A streaming box is a simple, slightly crude, method for upgrading a sluicebox concentrate. Such a unit has been found useful for alluvial tin deposits. It consists of a large container with a solid baffle board over which water runs.

Separation is a function of baffle angle and waterflow.

Operating information
The worker stands in the box and places the concentrate on the baffle board. Five to six shovelfuls are placed at the baffle base, thrown at the baffle, and the process repeated.

The baffle slope is found by experiment. During operation the tailings should be checked for mineral losses, and the baffle slope must be reduced if losses become high.

Source
Queensland Government Mining Journal, December 1977.
(Fabricate locally.)

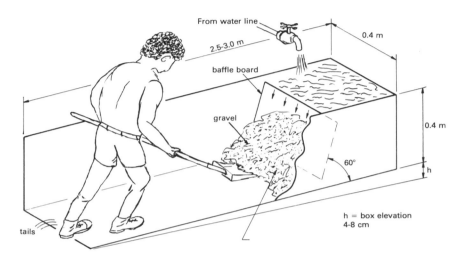

WOODEN SLUDGE BOX

Exploration drilling samples may consist of solid cores, sludge, or a mixture of both. The collection and assaying of sludge becomes particularly significant when there is insufficient core material. Fine sludge is particularly difficult to recover but it must all be retained, especially if the ore grade is marginal.

A sludge box is used to settle the particles. They are usually run in series, although the second box may collect only a small amount.

Operating information
When the sludge has settled, the clear water can either be run off through pluggable drain holes, or it can be siphoned off. Settling can be accelerated by the addition of lime, alum or flocculants.

Care should be taken to prevent fine loss. In order to avoid false results through contamination of the sample, the box must be cleaned thoroughly each time it is used.

Source
As used by the Cleveland Cliffs Iron Company, information taken from Peele, *Mining Engineer's Handbook*.

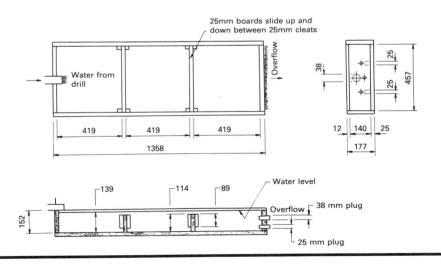

SLUICEBOX

The sluicebox or riffle box is a simple device designed to replace a batea (see Section 1). Higher throughputs can be obtained, allowing the processing of poorer quality alluvials. Separation results from the heavy mineral being retained behind the riffles during the flow of alluvium through the box.

There are many ways in which sluiceboxes can be set up. The illustration shows a simple arrangement using one box fed through one screen. Another common arrangement is to use a 25mm screen in front of the first sluicebox, with a 10mm screen placed between this box and a second.

Difficult clay lumps in the feed need to be broken up by a wooden tool, which is moved from side to side in the launder.

Fine gold may need further processing. Alternatively, the number of sluiceboxes can be increased or the second box may need to be wider.

Operating information

Recovery | 90 per cent of mined gold over 1.5mm in size, using two or three sluiceboxes.
Capacity | 2.3m³ of material per 10 hours, with a two-man operation

Water usage | 200 litres per minute, not recyclable. A 50mm pump should be used if there is insufficient head for gravity feeding.

Source
Illustration is based on a design in Fawcett, B., *UK Gold Prospector's Handbook*.

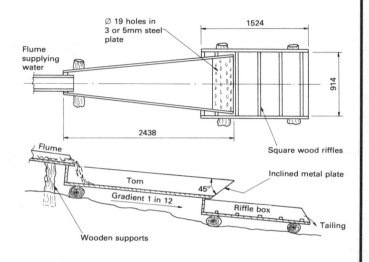

UNDERCURRENT

The undercurrent is a processing device used to improve the recovery of fine gold lost after a sluicebox.

The finer fractions that constitute the feed to the undercurrent are spread over a relatively wide area in a shallow stream. Riffles, at right angles to the flow, are used to retain the heavy minerals. A variety of materials can be used for the riffles.

The feed to the undercurrent from the sluicebox is arranged through a screen followed by a distribution box.

Operating information
The undercurrent is tilted at a gradient of between 8 and 12.5 per cent, the precise slope being found by experiment.

The heavy minerals are removed periodically from the riffles.

Source
Fabricate locally.

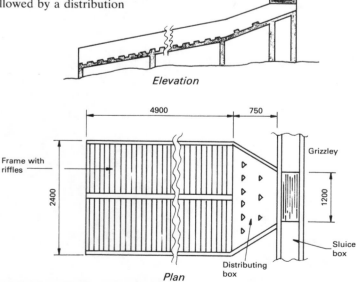

LUPA DRY BLOWER

This device is similar to the Mexican dry blower, but is well suited to the separation of coarse gold (for example, $+300\mu m$). It is also useful for tin, tantalite/columbite and microlite.

Power driven blowers force air through an inclined fine mesh screen which is set at an angle to the ore feed stream. The vigorous blast of air results in high throughputs.

Fine gold may be lost by such a system and some of this loss can be prevented by winnowing the undersize of the feed (see the diagram).

Operating information

Power	2 to 4 HP
Capacity	20 to 40m³ per day
Recovery	40-88 per cent
RPM needed	250 to 450

Sources
Manufacturers supplying dry separation equipment (not illustrated here) include: Fives-Cail Babcock (listed previously), and Joy Manufacturing Co., listed overleaf.

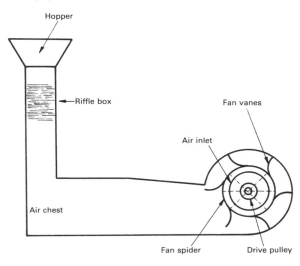

Principle of Lupa dry blower

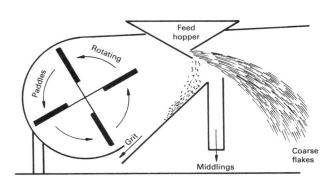

Winnowing principle

MEXICAN DRY WASHER

This washer can be used for dry placer deposits where water is scarce. If the conditions are good the results can compare favourably with wet washing.

A screen is used to remove all the oversize material before feeding into the hopper. The hopper then feeds a riffle board, which is suspended over manually operated bellows. The air from the bellows is used to fluidize the bed.

The ripple tray base is constructed of coarse wire netting, over which are laid one or two layers of burlap or thin cotton cloth, or a single layer of canvas.

Beneath the riffle tray the air supply is controlled by valves via an eccentric drive.

Operating information
Two men are needed to operate the washer (one to drive the bellows, three pumps of which result from each turn of the handwheel).

The washer must be levelled carefully before use and the feed material should be dry and well pulverized.

The canvas should be replaced after treating each 150 m³ of gravel.

Capacity	1.5 to 12 m³ per day, depending on the feed
Recovery	80 per cent, estimated.

Source
Fabricate locally.

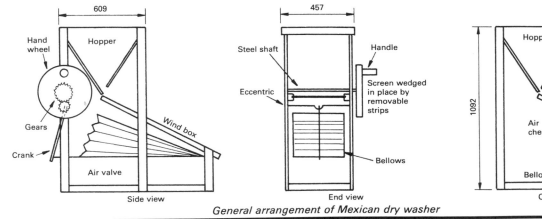

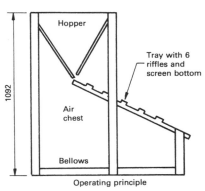

General arrangement of Mexican dry washer

DRY PINCHED SLUICE

The dry pinched sluice is the dry version of the sluicebox. It is simple and requires no moving parts or water. Instead of water as the transport medium, the feed is fluidized by an airflow introduced from beneath a porous bed.

The sluice consists of a tapered trough (wider at the free end). The porous trough base is mounted over an air box.

The coarse/heavy particles report to the lower part of the bed and a splitter at the discharge is used to make the final separation.

Such units have been used for separating pyrite from coal, cassiterite and phosphate, for secondary metal recovery and for dry tin alluvials in the Niger.

Operating information
The feed material should have a density difference greater than 1.0 gram per millilitre.

Capacity	1 tonne per hour for a sluice 750mm long × 200mm wide
	3 tonne per hour for a sluice 1500mm × 200mm
Feed size	minus 1mm
Air rate	1800ft³ per hour for every square foot of sluice.

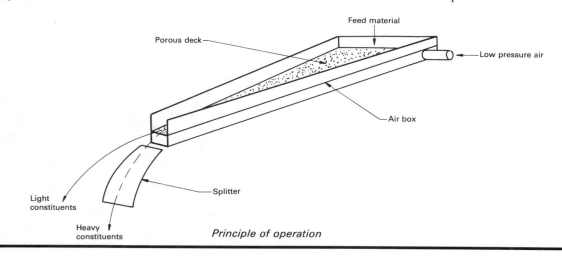

Principle of operation

TRAILER-MOUNTED FLOTATION MILL

Flotation mills have excellent recovery of both free milling and sulphide-type gold and silver ores. They are used to recover gold that is too fine to be recovered by gravity methods. They are faster than cyanide plants, and can selectively float different minerals such as gold, silver, copper, lead, zinc, tungsten, molybdenum, etc.

This example is one of a range of ready-to-run trailer-mounted units from this manufacturer, which also includes trommel wash plants, cyanide mills, crushing and grinding mills, jigs and table gravity mills. Capacities from 25 tonnes per day up to 500 tonnes per day are catered for.

Operating information
The manufacturer can carry out tests on ore samples at minimal cost, reporting on the recommended process and recovery results and sizing the tanks and flotation cells according to the mine's needs.

Flotation mills typically recover between 85 and 95 per cent of the gold and silver in normal ore types.

Basic specification
These mills are made in the following capacities:
Tonnes per day: 20, 50 to 75, 100, 200 and 500.
The following relate to the 50 to 75 tonne per day size:

Power	50 kW, 220 volts, three phase, 60 Hz
Weight	About 10 tonnes
Length	35 feet (10.7m)
Width	8 feet (2.44m)
Height	13 feet, 6 inches (4.1m)

Source
Sierra Mining and   Telephone 714 849 9789
  Manufacturing Co. Inc.
PO Box 1105
Banning
CA 92220
USA

Other sources

Outokumpen Oy.   Sala International AB
PO Box 280   PO Box 302
SF-00101 Helsinki   S-73300 Sala
Finland   Sweden

Joy Manufacturing Co.   Envirotech Corp.
1200 HW Oliver Building   (Wemco Division)
Pittsburgh   PO Box 15610
PA 15221   Sacramento
USA   CA 95813
                                   USA

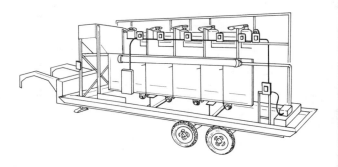

CONTINUOUS BELT FILTERS

The filters illustrated here are used in dewatering sludges and slurries. Previously manufactured by Mitchell Cotts, their production is now continued by Jenkins of Retford. They are made under licence from the Andriz Company, who give technical back-up support for the product.

The principle is a continuous permeable belt, driven over rollers, through which water is extracted with pressure assistance.

Operating information

These filters are made in a range of belt widths and capacities in two styles, as illustrated.

The company will arrange for laboratory testing to advise on the specification best suited to particular requirements. A mobile hire plant, with a 1 metre wide belt, is available for pre-installation trials (where the mine location is suitable).

Belt widths from 1 metre up to 3.5 metres.

Belt speeds do not usually exceed 12 metres per minute.

All essential parts are corrosion resistant or protected.

Service life is claimed to be up to 15 years of three shift working, with minimal maintenance.

Typical performance figures are:

	Coal flotation waste	Magnesite
Dry solids		
In slurry	25 to 35	20 to 25 per cent by weight
In cake	56 to 38	70 to 76 per cent by weight
Throughput		
Slurry	40 to 60	25 to 50 cubic metres/hour
Dry solids	12 to 20	6 to 12 tonnes per hour

Source

Jenkins of Retford Ltd
Thrumpton Lane
Retford
Nottinghamshire
DN22 7AN
UK

Telephone 0777 706777
Telex 561222

Other sources

Larox
Box 29, SF-53101
Lappeenranta
Finland

Eimco Process Equipment
PO Box 300
Salt Lake City
UT 84110-0300
USA

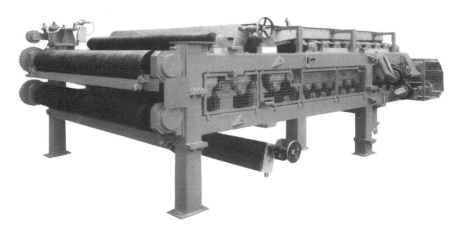

Belt filter type SMX

Belt filter type S7

TRANSPORTABLE GOLD PLANT

Designed for easy mobility, and capable of being used in remote areas, this single plant brings together in six modules all the elements of gold processing needed to produce the end product.

The basic flowsheet assumes a readily accessible oxidized ore, which is easily mineable with minimum blasting and amenable to cyanide leaching and carbon-in-pulp adsorption.

The design allows flexibility within the individual process elements chosen. The modules cover the following operations:
— Crushing and screening
— Grinding and classification
— Leaching
— Adsorption
— Elution, electrowinning, gold recovery and carbon regeneration
— Services and reagents.

Operating information

Capacity	20 tonnes per hour
Feed size	Minus 200mm into crusher
Discharge size	Minus 50mm from jaw crusher
	Minus 12mm from core crusher
Grind size	75 per cent minus 74 μm
SG feed solids	2.5
Work index	10 to 13 kWh/t

Dimensions when transporting:

— Height	4.5m
— Length	15m
— Width	4.5m
— Weight	75 tonnes

Source

Jenkins of Retford Ltd	Telephone 0777 706 777
Thrumpton Lane	Telex 56122
Retford	
Notts	
DN22 7AN	
UK	

For other manufacturers of mobile filtration equipment, see sources listed under **continuous belt filters.**

TRANSPORTABLE HEAVY MEDIA PLANT

Here is a heavy media plant that can readily be taken apart and packaged for long distance transport. It is designed to perform gravity concentration of minerals with sufficient gravity differential between the desired value and the host rock. The plant is intended for the preconcentration of ore prior to further liberation and additional processing (such as flotation).

The plant design can be adapted to suit the mineral to be treated, and modules can be added or subtracted to suit both geological and metallurgical conditions.

Typical minerals processed are diamonds and alusite, but others are suitable if the specific gravity differential is suitable and the ore can be liberated.

Operating information

Media	Ferrosilicon or magnetite (depending on ore)
Feed size	From 30mm to 0.5mm
Capacity	20 tonnes per hour
Power	143kW
Water needed	70m³ per hour
Height	4.6m
Length	7.6m
Width	4.15m

Source

Jenkins of Retford Ltd	Telephone 0777 706 777
Thrumpton Lane	Telex 56122
Retford	
Notts	
DN22 7AN	
UK	

PORTABLE CRUSHING AND SCREENING PLANT

A chassis constructed of steel channel and other rolled steel members of substantial section is used to contain this crushing and screening plant. The chassis is mounted on pneumatic wheels, and fitted with a drawbar.

A jaw crusher, with wearing surfaces made from manganese steel, is belt-driven from a diesel engine. The crusher can be supplied as a separate unit, and can be fixed or mounted on a truck as a mobile unit.

5mm steel plate is used for the underslung rotary screen, which can be provided punched with holes in two standard alternative patterns or to special order.

Operating information

Jaw crusher:

Action	Single toggle, with roller bearings
Mouth size	508 × 279mm
Settings	18mm to 64mm
Speed	340 to 350 rpm
Capacity	18 to 25 tonnes per hour
Power	22 HP electric motor or diesel engine of between 27 and 40 HP driving a flat belt

Rotary screen:

Construction	5mm steel plate mounted on a 64mm diameter shaft by cast iron stays
Power	Through bevel gears from a pulley driven by a belt from the flywheel side of the crusher spindle
Dimensions	2.45m long × 690mm diameter
Sections	The screen is divided into three gratings of 915mm, 762mm and 762mm length, with 690mm overlapped by wire mesh on the intake side
Punchings	25mm, 37mm, 50mm or 64mm

Source

Sur Iron and Steel
Co. Ltd
15 Convent Road
Calcutta 14
India

Telephones 24-1307
24-1308

The jaw crusher unit

The complete mobile unit, comprising jaw crusher, rotary screen, wheeled chassis, diesel engine and drives

MOBILE CONCENTRATING TABLE

The table described here is known as the standard half size. Mounted on a strong steel chassis, it is light enough to be towed even by a family saloon car, although a four-wheel drive Land Rover-type of vehicle is obviously more suited to rough terrain.

This is a mobile adaptation of a fixed unit, and the 2 HP electric motor has been replaced by a small petrol engine (a diesel alternative is available) to allow operation in remote places.

Tables such as this are used widely for a variety of heavy minerals. Tables are particularly useful for washing ore which is unsuitable for flotation.

Concentration takes place when a water/solids mixture is allowed to flow over the surface so that the heavier particles sink before the lighter particles. The motion induced by the motor causes the heavier particles to move to one end of the table, whilst the lighter tailings are washed over the lower edge.

A suction pump can be supplied as an optional extra. A smaller mobile laboratory size table is also made by the same company.

Operating information

Capacity	250 to 500 Kg per hour
Table size	2473 × 1219mm
Overall length	4300mm
Overall width	1400mm
Road weight	1070 Kg

Source

Wilfley Mining   Telephone 0933 226368
  Machinery   Telex   317220
Cambridge Street
Wellingborough
Northamptonshire
NN8 1DW
UK

LIGHTWEIGHT GOLD DREDGING UNIT

This is a lightweight, highly manoeuverable alluvial gold dredge. It is specially adapted to operate on rivers when rocks and tree trunks form frequent obstructions, where rapids are numerous and where water levels can be very low for several months of the year.

The device comprises a pontoon, made of 0.5mm wide cubic structures of plastic and fibreglass, carrying a gravel pump and gravel treatment unit.

Water, pumped under pressure towards the top of the suction hose, induces a venturi effect which draws gravel into the tube. Gravel feeds onto a sluice, where heavy minerals are trapped by riffles. Gold can be recovered by gravity concentration (panning) or by amalgamation.

Operating information

The unit can be operated by one man, plus a diver to shift the suction head.

Suction hose	6 inch (150mm) diameter, semi-rigid
Capacity	2 to 4m³ of gravel per hour
Weight	1.2 tonnes

Maintenance and repairs can be handled locally. Units have been installed in Africa (Ivory Coast, Mali, Guinea and Cameroon).

Source

Bureau de Recherches   Telephone 38 64 33 45
  Geologique et Minieres
  (BRGM)
BP 6009
F-45060 Orleans Cedex 2
France

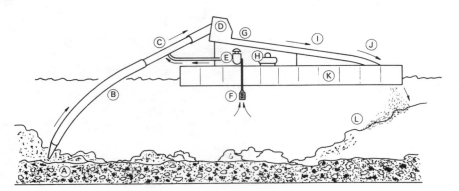

A. Gold-bearing gravels.
B. Semi-rigid tube, 150mm diameter.
C. Suction hose.
D. Caisson.
E. Motor and pump.
F. Filter.
G. Pre-concentration of gold and heavy miners.
H. Dry air compressor.
I. Riffles.
J. Sluice.
K. Pontoon.
L. Treated gravel.

SUCTION DREDGE

A small one-man machine for sucking deposits up from riverbeds. The dredge consists of a raft, sluicebox, engine, pump and suction pump. A standard car inner tube is used for flotation.

If the recovery is to be carried out at a depth greater than 1m, a compressor can be added to supply a diver with air.

The unit can be broken down into easily portable sections.

Operating information
Assembled weight   18Kg

Maximum capacity	1.8m³/hour
Fuel consumption	0.2 gallons/hour
Depth of operation	3.05m
Suction hose diameter	50mm

The unit can be operated by one man.

Periodically, the heavy minerals must be removed from the sluicebox and further treated to obtain the desired concentration.

Source
Keene Engineering Inc.   Telephone 818 993 0411
9330 Corbin Avenue   Telex   662617
Northridge
California 91324
USA

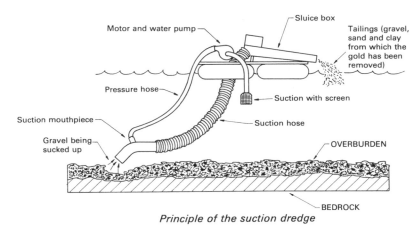

Principle of the suction dredge

JAW CRUSHER

Jaw crushers are intended for the coarse crushing of hard, fracturable ores or other rocks. They are usually used as primary crushers.

The crusher has one moving plate and one stationary. Coarse rock is fed into the top, being crushed progressively as it moves downwards. The drive is from a flywheel-assisted electric motor through a belt.

Crushers are available in a wide range of sizes, from laboratory and pilot plant up to large production units. A general rule when choosing a crusher is that the opening should accommodate a 70 to 75 per cent top size of rock.

Operating information
An example from this manufacturer's range:

Power	Electric motor, 5 HP
Inlet opening	101 × 152mm
Batch capacity	Approx 20kg in 5 minutes (for moderately hard quartz)
Weight	338 Kg

Source
Sepor Inc.   Telephone 213 830 6601
718N Fries Avenue   Telex   194456
PO Box 578
Wilmington
CA 90748
USA

Other sources
Other manufacturers supplying laboratory mineral processing equipment are: KHD Humbolt Wedog, Pegson Ltd and Envirotech Corp. (Wemco Division), all previously listed. Also:

Siebtechnik GmbH
Postfach 101751
D-4330 Mannheim (Ruhr) 1
West Germany

Denver Equipment Division
 (Joy Process Equipment Ltd)
9 North Street
Leatherhead
Surrey
KT22 7AY
UK

DOUBLE ROLLS CRUSHER

Double rolls crushers are used for controlling product size with the minimum generation of fines. The crushing action is derived from two contra-rotating rolls.

Rolls crushers are specially good for crushing fracturable substances from about 64mm to 12-20 mesh. Generally the reduction ratio obtainable is 3 to 1, but 4 to 1 is possible with softer rocks and lower capacities.

Machine bearings are spring loaded to reduce the wear effect of heavy impact. The surfaces of the rolls are hardened alloy.

Before entering the crusher, the ore is fed over a grizzley and discharged into a hopper. An automatic feeder is available which allows the crusher to be integrated into a continuous process.

Operating information
Example of this manufacturer's range:

Batch capacity	10-20 kg approx every 10 minutes
Power	2 Hp
Drive	Chain
Diameter	203mm
Width	127mm
Weight	386 Kg

Capacity depends greatly on rock hardness and machine setting.

Source
Sepor Inc.
718N Fries Avenue
PO Box 578
Wilmington
CA 90748
USA

Telephone 213 830 6601
Telex 194456

LIGHTWEIGHT PORTABLE MILL

This portable mill weighs only 17 lbs (under 8 Kg) but can crush ore up to 25mm size. It is intended as a tool for prospectors and geologists, and offers a considerable advantage over the use of a pestle and mortar. Hard rock samples can be pulverized in a few seconds, greatly increasing the amount of field work that can be covered.

The mill casing is aluminium lined with stainless steel and the drive is taken directly from a petrol engine without intervening gears or belts. Two ordinary chain shackles are used as rotating hammers. All wearing parts are easily and economically replaceable.

The discharge can be blanked off for batch testing, or interchangeable discharge screens can be fitted. The standard screen will reduce rock to a 60-80 mesh.

A carrying handle is fitted which also serves as the engine throttle control.

(NB: Not to be used on free-milling gold ores.)

Operating information

Type	Goldwind Pocketmill
Power	Integral Echo petrol engine
Capacity	Rock sized up to 25mm
Speed	Hardrock to fine grind takes a few seconds
Weight	17 lbs (7.71 Kg)

Source
The mill is manufactured by Goldwind Inc, but our information was obtained from:
Interex Development Corp.
1003-470 Granville Street
Vancouver BC
V6C 1V5
Canada

Telephone 604 688-4155

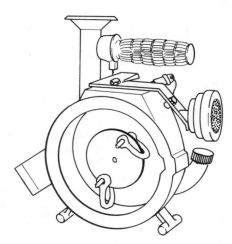

The mill shown with its quick release cover removed. The stainless liner slides out easily, allowing discharge screens to be changed.

STACK OF HAND-OPERATED SCREENS

The use of a stack of screens, arranged in order of decreasing mesh size, allows a complete size analysis of mineral prospects and deposits to be undertaken.

The entire stack is placed over a rocker (made from conveyor belting) and the whole assembly is mounted on a sturdy wooden base.

The rocker serves two purposes:
1. It is used to impart vibration to the stack, ensuring that all the material is screened.
2. It acts as an enclosed collecting pan for all the undersized material which has passed through the bottom screen of the stack.

The bottom screen has provision for a handle, to allow the stack to be vibrated by foot.

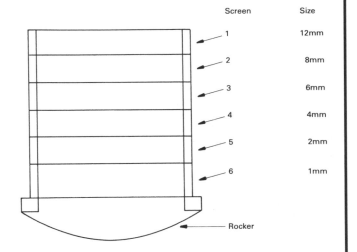

Screen	Size
1	12mm
2	8mm
3	6mm
4	4mm
5	2mm
6	1mm

Rocker

SMALL HAND-HELD VIBRATING SIEVE

This is a very small hand-held vibrating sieve for dry or wet screen particle size analysis of small samples.

It is used either as a single screen, or with multi-deck sieves. Accessories available include a holder (for wet or dry screening), feeding equipment, receiving pan, support and other items.

Operating information
Typical screening time 5 to 10 minutes
Vibrator frequency 100 Hz
Power 24 volts, 50 Hz, single phase, 50 Watts. 110 volt and 240 volt options are available.

Screen diameter 200mm or 100mm
Mesh 140, 160, 200, 240, 280, 300, 320, 360, 400 and 500
Number of decks Up to 5
Weight of vibrator 1.8 Kg

Source
Liuzhou Mineral Telephone 2704
  Prospecting Machinery Telegram 2232
  Works
West Yen An Road
Liuzhou
People's Republic of China

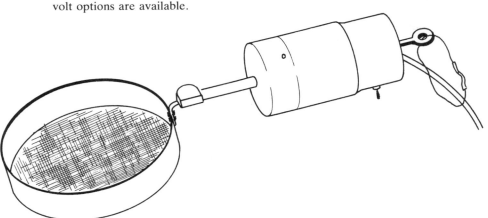

5. Haulage, handling and transport

The selection of equipment illustrated in this section is not comprehensive but it covers the basic requirements for collection, control, loading, transportation and hoisting of ore (product), waste and materials between underground and surface.

The criteria governing the choice of equipment relate to a number of factors:
— geographic location
— the state of industrial development within the country
— the economic situation
— the mining method to be used
— the configuration of the particular mine, i.e. access by adit, ramp, decline or vertical shaft.

As the individual activities of collection, control, loading and transportation are interdependent it is necessary to consider the equipment used to perform these functions as parts of a system whose performance must be matched to one another. For example there is no point in having a railway system capable of transporting 200 t/hour if the loading capability is only 50 t/hour.

Many inexperienced owners/operators fall into the trap of undersizing equipment, or having an insufficient number of units, to meet production targets adequately. This is because insufficient consideration is given to estimating realistically the downtime for routine maintenance, breakdowns and general operation difficulties which will be experienced. This is discussed further below under *Equipment and machinery*.

Before the purchase of any item of equipment is made the capital cost is invariably given serious consideration. However, there are many other points that are at least as important. The major ones are:

(a) Operating costs: these include fuel or power, tyres (where appropriate), spare parts and operating labour.

(b) Ease of operation: is the item 'user friendly'? Does it impose any physical or mental stress on the operator as a result of awkward controls, noise, or similar conditions?

(c) Maintainability: is there a high maintenance requirement because of rapidly wearing parts? Are spares readily available locally? Is there a local agent? Can it be maintained by the existing maintenance crew or are specialists needed?

(d) Mining and machinery regulations: it is, of course, mandatory that any item of machinery or equipment conforms with the local mining and machinery regulations and meets all health and safety requirements. If there is any doubt it would be prudent to discuss the matter with the local inspectorate.

Equipment and machinery

In general it is always good policy to purchase simple, rugged equipment and machinery which is easy to operate and maintain, providing it is adequate for the job. It should also be borne in mind that complex machines and equipment usually need higher skilled operators and maintenance staff, more sophisticated service tools and better workshop facilities.

Collection and control

These are some of the items of equipment which are used for the movement and control of broken ground (product and waste) within a stope or at a face. The requirement is to gather and move the broken ground in a safe and efficient manner to an ore pass or to a pile for subsequent handling.

A typical installation would be a slusher/scraper haulage in a 'flattish' stope scraping ore to a pass. Control of the ore from the pass could be by a stope box or door and chute arrangement for subsequent loading of rail cars (wagons). The rate of production from a scraper depends mainly on the horse power of the unit and advice should be obtained from suppliers on this aspect.

Stope boxes and control doors and chutes are usually made locally with designs being developed to suit local conditions.

Loading

The two LHDs (load-haul-dump units) and the rocker shovel are specifically designed for underground production applications while the mini and multi loaders are general-purpose machines more suited to clean up operations and such like.

The LHDs illustrated are the smallest units currently available and because of their small width dimension are particularly suited for working in narrow openings. LHDs are manufactured in a wide range of engine and bucket capacities and overall sizes. It is therefore worth while for operators to discuss their requirements with suppliers before selecting a particular machine.

The choice between electric and diesel prime mover will depend upon a variety of factors not the least being the availability of sufficient electric power underground. Two major advantages of the electric powered unit over the diesel are there are no noxious fumes (therefore extra ventilation is not required) and the noise level is significantly lower.

An electric unit will however be restricted in its operating area by the length of its electric cable while the diesel unit has no such restriction.

The maintenance cost and effort to maintain LHDs in a good state of repair is significant and depending upon the number of shifts worked per week the overall availability could be around 60 to 70 per cent.

The rocker shovel is a very popular machine worldwide which is used extensively in tunnelling operations. It is very robust and comparatively simple to operate and maintain.

Like the LHDs these units come in a variety of sizes and capacities and again it is advisable to discuss particular needs with two or three suppliers before making a final decision.

Underground transportation

The efficient operation of any underground mine depends largely on the underground transportation system. The ability to move ore and waste quickly to surface and to move supplies in the reverse direction is often the determining factor between profit and loss.

The most popular method over the years particularly for small mines has been a rail system using cars and locomotives (and is used worldwide in mines ranging from the smallest to the largest). However many mines now have integrated systems which use a combination of conveyors, tracked and rubber-tyred equipment.

Although the cars illustrated in this section are basic designs and are only indicative of the type of units available, they are well matched to the stope box and chute discussed under Collection and control.

As transportation is the link between the production areas and hoisting it is obvious the designs and production schedules must be compatible.

It is essential to be conservative in calculating the number of cars, locos and spare batteries required to ensure continuous operation. The calculation which is fairly simple includes loading time, travelling time, time to discharge and return. Due allowances should be made for unforeseen hold ups such as lack of ore, breakdowns, too much ore at the shaft bottom, changing batteries, track problems and maintenance.

Hoisting

The hoisting elements are shaft steelwork, loading arrangements, conveyances, ropes, headframe and winders. Air and electric winches are really general purpose units which under normal circumstances would not be used for a hoisting duty. However with modifications they could be used to sink shallow shafts or as hoists in a service shaft for handling materials.

One of the illustrations shows a hoist for small shafts which is a self-contained unit complete with headframe. This is probably adequate for a shallow shaft producing a very small tonnage, or for mining a shallow exploration shaft and potential users should investigate its limitations thoroughly before purchase.

Hoisting shafts can be either inclined or vertical and although there are slight variations in the regulations governing their operations safety is always of paramount importance. The regulations covering the whole area of shafts and hoisting are very strict, especially for man winding, and owners and operators must make themselves familiar with them.

Depending upon the angle of an inclined shaft an alternative to installing a winder is a conveyor, although the maximum angle for this to operate satisfactorily without too much spillage is about 15 degrees. Its use is also limited by the lump size of the rock and this should be investigated thoroughly before considering this option as it may be necessary to primary crush the rock underground prior to feeding the conveyor.

In general the design, sinking, equipping and winder installation of any steeply inclined or vertical shaft, unless of a small cross section and shallow depth, should be carried out by engineers specializing in this field.

Although the companies listed in this section are equipment suppliers almost all of them would be capable of putting together a package with other specialists to carry out a total project.

The choice of ropes for particular applications is also an area where specialist advice should be sought. This route should ensure that the correct type of rope and material are being used from the safety angle, and also ensure maximum rope life which can have a significant effect on costs.

Ernest Hogg

MINI LOADER

Although ruggedly constructed from heavy duty box section members, this self-propelled mini loader is extremely light. It can also be towed behind a vehicle.

The bucket motions are controlled by hydraulics, the pressure for which is provided by a pump powered from the petrol engine.

Excellent manoeuverability is demonstrated by the fact that this machine can turn in its own length.

Underground applications for the machine are likely to be limited, as mining regulations in some countries prohibit the use of petrol underground. It is probably of most use for 'clean up' operations on the surface, rather than as a production tool.

Operating information
Hopper capacity	8 ft³ (0.226 m³)
Power	Petrol engine, 5 HP
Pump	15 litres per minute
Maximum speed	5 km per hour
Length	2134mm (84 inches)
Height	1068mm (42 inches)
Width	890mm (35 inches)
Weight	290 kg

Source

Team Services International Ltd	Telephone 0536 711246
Kyngswoode	Telex 341543
Rushton Road	
Rothwell	
Northants	
UK	

SMALL MULTI-LOADER VEHICLE

The small size and variety of attachments of this multi-purpose materials-handling device makes it extremely manoeuverable and adaptable. However it is not designed as a production unit and is likely to find most use at surface, in small mining operations.

The system is hydraulic, powered from a diesel or petrol engine. The operator stands at the back, giving good visibility, and the controls have a deadman's handle safety feature.

Many optional attachments are available for this machine, and include:
— Bucket
— Borer
— Dozer blade
— Hydraulic breaker
— Ripper
— Leveller
— Scraper

Operating information
Standard bucket
— Width	930mm
— Capacity	0.1m³

Performance with standard bucket
— Digging depth	1.7m maximum
— Reach	2.5m
— Load-over height	1.8m
Load capacity	225 kg maximum
Forward speed	5 km per hour maximum

Power	Honda 10 HP petrol engine or Lombardini 8.5 HP diesel
Subframe slew	125 degrees
Wheel width	127mm
Weight	570 kg

Source

Opico Ltd	Telephone 0778 421111
South Road	Telex 32637
Bourne	
Lincs	
DE10 9LG	
UK	

SLUSHER/SCRAPER HAULAGE

This is a simple system which is used to scrape material to a control point. Extensively used in underground 'flat' stopes to move primary broken ground to ore-passes in medium tonnage operations.

The slusher system is a mechanized mineral haulage method which provides a simple and extremely rugged link between the point of breakage and main haulage. The system has a high loading rate, eases loading, and attracts very low maintenance costs. It can be used also for mineral dumps. Such units are used extensively, worldwide, in a variety of mines.

The system comprises:
— A prime mover
— Ropes
— Return pulley
— Scraper

The scraper bucket digs into the material to be loaded. The full bucket is then pulled along the floor to a discharge point. The scraping distance can be increased progressively by repositioning the return pulley.

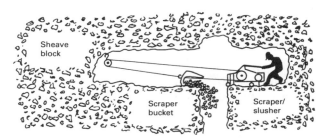

Scraper installation in underground stope

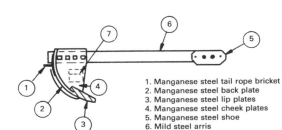

1. Manganese steel tail rope bricket
2. Manganese steel back plate
3. Manganese steel lip plates
4. Manganese steel cheek plates
5. Manganese steel shoe
6. Mild steel arris
7. Turn over bracket

Operating information

Scraper shape	Varies according to the task, capacity, loading rate and the mineral to be handled
Return pulleys	3 or 5 tonne capacity, with sealed bearings
Drive	Compressed air, or electric (flameproof or non-flameproof as appropriate)
Drive example	Non-flameproof electric motor, 15 HP at 1500 RPM
Rope speed	68 to 76m per minute
Rope capacity	84m of 13mm diameter wire rope

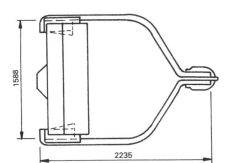

Typical scraper bucket

Source

Pikrose & Co. Ltd   Telephone 061-370 1368
Delta Works   Telex   667488
Delta Road
Audenshaw
Manchester
UK

Other sources

Joy Manufacturing Co.
1200 HW Oliver Building
Pittsburgh
PA 15221
USA

Eimco
 (Great Britain) Ltd
Earlsway
Team Valley
Gateshead
NE11 0SB
UK

Atlas Copco MCT
S-104 84 Stockholm
Sweden

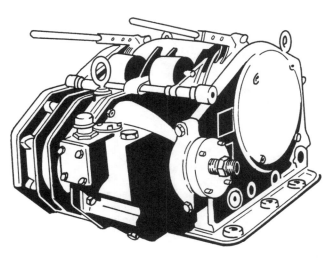

Drive unit

CHAIN CONVEYOR

In underground applications, chain conveyors are used mainly in coal mines and in particular to remove broken coal from 'long wall' faces. When used in a 'long wall' system the conveyor is only one part, and as the complete system is costly, these systems are normally only installed in larger mines.

This is a short, relatively low powered but flexible conveyor. It can be used to mechanize the movement of material over short distances, whether on the surface, or within a plant, or underground from the working place to the transportation system.

The drive head at one end of the chain comprises a motor, gearbox and coupling, and the chain is fed back over a sprocket at the other end.

The sectional construction of the conveyor allows it to be snaked or bent by rams, and it can thus be advanced into the next cut in sections. Applications can also extend to other mineral workings.

Operating information

Capacity	40 to 100 tonnes per hour
Power requirement	15 HP
Chain centres	300mm
Chain size	10 or 14mm

Source
Wultex Machine Co.   Telephone 0602 302603
 c/o Becorit Ltd   Telex   37526
Hallam Fields Road
Ilkeston
Derbyshire
DE7 4BS
UK

Other sources
Dowty Mining
 Equipment Ltd
Ashchurch
Tewkesbury
Gloucestershire
GL10 6JR
UK

Harwood
Team Valley Trading Estate
Gateshead
NE11 0LP
UK

Joy Manufacturing Co.
1200 HW Oliver Building
Pittsburgh
PA 15221
USA

Roxon
Keskirkankaantie 9
SF-15860 Hallola
Finland

BELT STAGE LOADER

This type of machine is of use in high-speed tunnelling of large headings where mining is continuous.

This particular unit is designed for use in underground driveway operations with a section area of more than five square metres. It is used to transport rock from a scraper loaded into mine cars.

The drive is a flameproof oil-cooled electric drum, resulting in easy assembly and dismantling.

Operating information

Motor power	10 kW
Voltage	380/660 volts
Belt speed	1.25 metres per second
Belt width	650mm
Belt length	12.3m plus an extension option
Load per operation	6 to 8 one-tonne mine cars can be loaded in each conveying operation
Working capacity	100m³ per hour
Rail gauge	600mm or 900mm
Minimum track radius	20m
Length overall	16.8m or 21m
Width	1.5m or 1.9m
Height	1.8m
Weight	500 kg

Source
China Coal Mining   Telephone 464847, 464010
 Machinery Manufacture Cable   6671
 Corp.
21 He Pingli Bei Street
Beijing
People's Republic
 of China

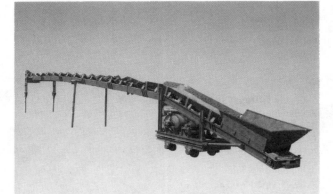

HYDRAULIC RAM

Hydraulic rams are used wherever a large force is required, for example: moving chain conveyors and roof chocks, activating chute doors, finger controls and wagon tipplers. This is a double-acting hydraulic ram. The ram must be anchored securely in order to withstand the reaction force. A prop can be used or, when pulling, a secured chain.

A ram such as this could be used for other purposes, such as pushing wagons and so on. However, compressed air operation is usually preferred in those circumstances.

This unit may be fitted with leather seals and valve seats which allow the ram to be used even with badly damaged piston rods and cylinder bore surfaces. The valve seat and disc are made from rust-resisting steel and can be replaced quickly.

Operating information

Stroke	760mm
Thrust	3.15 tonnes at 70 kg per cm^2
Pull	1.75 tonnes at 70 kg per cm^2
Hose connections	16mm BSP pressure feed
	25mm BSP return
	Reducers or blanking plugs can be fitted to suit any system
Length	1180mm (closed)
Weight	58 kg

Source

Dowty Mining   Telephone 0684 292441
 Equipment Ltd   Telex   43285
Ashchurch
Tewkesbury
Gloucestershire
GL20 8JR
UK

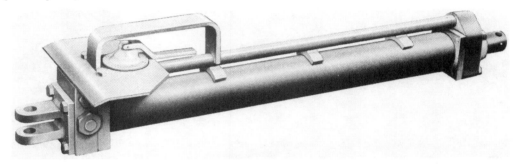

COUSIN JACK STOPE BOX

The Cousin Jack type box is one of the simplest methods used to control the transfer of broken ground from ore/waste passes to rail cars or similar transport, and is suitable for use in small, labour-intensive mines.

It comprises a fixed wooden box chute, the bottom surface of which can be reinforced if necessary to reduce wear by lining it with iron (old rails, for example).

Operating information

Stop boards are dropped behind iron straps to dam and control the outflow.

Source

Fabricate locally

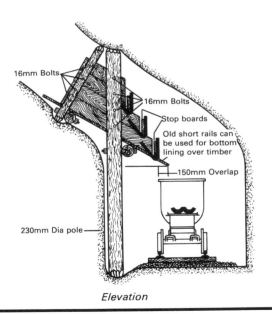

16mm Bolts
16mm Bolts
Stop boards
Old short rails can be used for bottom lining over timber
150mm Overlap
230mm Dia pole

Elevation

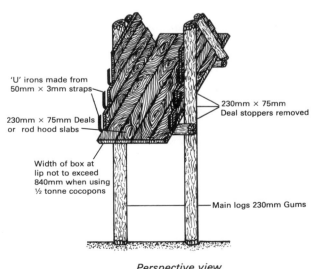

'U' irons made from 50mm × 3mm straps
230mm × 75mm Deals or rod hood slabs
Width of box at lip not to exceed 840mm when using ½ tonne cocopons
230mm × 75mm Deal stoppers removed
Main logs 230mm Gums

Perspective view

DOOR AND CHUTE ARRANGEMENT FOR ORE BIN DISCHARGE

This chute is designed to feed from an ore bin to another transport system, such as mine cars or hoists. However, this type of chute could also be used in similar applications to the 'Cousin Jack' box. The illustration shows manually operated doors, but for large installations it is common practice to incorporate compressed air cylinders (rams), with remote control.

Various arrangements include the following:

Conventional door
This is the most common design, with a swivelling single door located at the bin outlet (top end of the chute).

Cut-off door
A swivelling single door is arranged so as to cut off the flow at the discharge end of the chute.

Double door
Using both the above types of door in combination allows a fixed amount of ore to be discharged from the chute each time, the amount depending on the volume of chute bounded by the two doors (this might be arranged as one hoist load, for example).

Operating information
The chute is constructed from 6mm steel plate and angle iron. The dimensions will depend on the amount of ore to be discharged at a time.

Typical dimensions:
Ore bin opening   450mm × 610mm
Length   1.14m

Source
Fabricate locally

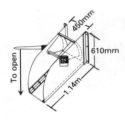

Conventional single door

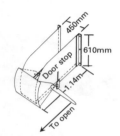

Single cut-off door for discharge end

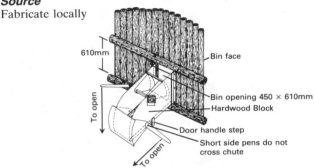

View of bin with chute fitted, using both door types in a double-door arrangement

ROCKER SHOVEL

This type of loader is fairly simple, robust, and depending upon the particular model, operates on rail tracks, has steel banded wheels for 'free' operation or alternatively crawler tracks. It can be used for most loading operations, but probably its most common use is in tunnelling. The loaded bucket is lifted over the top of the machine, enabling it to be operated in places where width is restricted. It discharges at the rear, normally into rail cars. One man is required for operation, and it is used in all sizes of mine. With slight adaptation, work can be carried out on gradients of up to 1 in 2.

The machine is used extensively throughout the world in all types of underground mine, but of course requires a supply of high pressure compressed air.

Operating information
Capacity	0.34 to 1m³
Loading capacity	0.5 to 1m³, average per minute
Bucket capacity	0.13 to 0.17m³
Weight	2.04 tonnes
Width	0.864m
Length	1.803 to 1.880m
Discharge height	1.168 to 1.500m
Air consumption	7.08m³ per minute
Air pressure	4.2 to 7 kg/cm²

Source
Eimco Mining Machinery   Telephone 091 487 7241
  International   Telex   53482
Team Valley
Gateshead
NE11 0SB
UK

Other sources
Atlas Copco MCT
S-104 84
Stockholm
Sweden

Saltzgitter Machinen
  und Autogen
Postf. 51 16 40
D-3320
Saltzgitter 51
West Germany

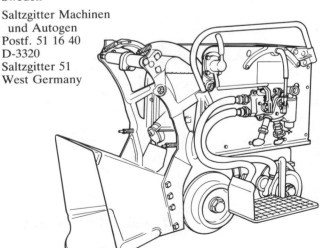

ELECTRIC LOAD-HAUL-DUMP UNIT

Load-haul-dump units are used extensively as production units and also for development work in drifts.

The model illustrated is an electrically-driven load-haul-pump unit, particularly useful for narrow veins and selective mining. It is small, fast and highly manoeuverable. As the name suggests, this type of machine picks up a load, transports it (within the limits of its cable), and dumps into an ore/waste pass or other transfer system.

The automatic cable reel at the rear can handle 85m of cable and the unit can traverse over 170m.

Electric drives do not create pollution, and produce little heat or noise.

The operator is located away from the loading operation and also has good visibility.

Operating information

Loading capacity	17 to 20 tonnes per hour, at 50m of travel
Bucket capacity	0.385m³
Tramming capacity	600 kg
Speed	0 to 7.2 km per hour, on level grades
Electric supply	380/440/525 volts, 50/60 Hz, 3 phase
Maximum bucket height	2m
Height	1.9m, with operator
Length	4.195m
Width	0.85m

Source
France Loader   Telephone (1) 45 01 81 24
50 Avenue Victor Hugo   Telex   610713
75116 Paris
France

Other sources
Wagner Mining   Atlas Copco MCT
 Equipment Co.   S-104 84
PO Box 20307   Stockholm
Portland   Sweden
Oregon 97220
USA

Eimco (Great Britain) Ltd
Earlsway, Team Valley
Gateshead
NE11 0SB
UK

DIESEL LOAD-HAUL-DUMP UNIT

The compact size and manoeuverability of this load-haul-dump machine makes it very suitable for narrow veins. It is also suitable for deposits requiring selective mining, minimizing waste dilution.

The LHD illustrated here is powered by a reliable and clean burning diesel engine. The advantage of this is that there is no need for an electric power supply, but the disadvantage is the need for additional mine ventilation.

The operator sits in the middle of the machine, facing sideways. This makes effective use of space, and allows good visibility when driving in either direction.

Operating information

Bucket capacity	0.4m³
Tramming capacity	680 kg
Speed	9.6 km per hour, level grade
	3.7 km per hour, 30 per cent grade
Power	Deutz F2L-511W diesel engine, 29 HP at 2800 rpm
Height	1.725m, with operator
Length	3.63m
Width	0.81m
Turning radius	1.37m inside
	2.49m outside

The ventilation rate required for exhaust dilution is 70m³ per minute.

Source
DUX Machinery   Telephone 514 581 8341
 Corp.   Telex   05-828859
PO Box 80
Repetigny
Montreal
J6A 5H7
Canada

Other sources
Eimco (Great Britain) Ltd   Toro (YIT)
(As above)   PO Box 434
                              SF-20101 Turku 10
Wagner Mining   Finland
 Equipment Co.
(As above)

Schapf Maschinenbau GmbH
PO Box 750360
D-7000 Stuttgart 75
West Germany

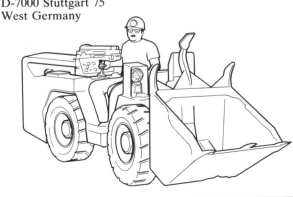

BUCKET TRUCK

This bucket truck has a chassis specially designed to carry shaft buckets (kibbles). This considerably reduces material handling, as the same bucket wound from the surface can be put on a wagon and hauled into the mine — or vice versa.

Commercially available wheels and axles are available (see the separate entry). The chassis is made from deal, braced with iron straps.

Use of this type of conveyance for regular hoisting will depend upon local mining regulations and is applicable for small tonnage operations.

Operating information

The underground rail system must be extended to run under the vertical shaft.

Dimensions given on the drawing are for guidance only, and will have to be chosen to suit the track gauge.

Care must be taken that buckets do not overhang the truck, and will align properly when being loaded, to ensure that the truck does not become unstable and tip over.

Source

Manufacture locally to suit individual specifications.

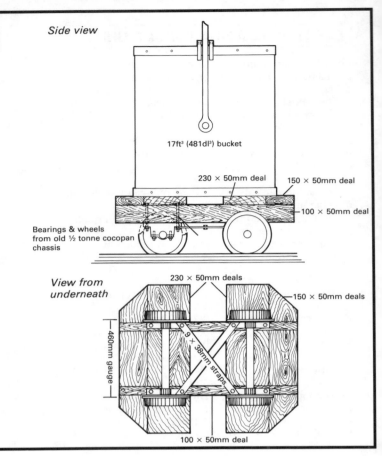

Side view

17ft³ (481dl³) bucket

230 × 50mm deal

150 × 50mm deal

100 × 50mm deal

Bearings & wheels from old ½ tonne cocopan chassis

View from underneath

230 × 50mm deals

150 × 50mm deals

460mm gauge

9 × 38mm straps

100 × 50mm deal

MINE CAR

Mine cars are the most versatile rail vehicles available. They can be used for carrying the product, waste, and for materials in general. Cars can be hauled by many methods, which include ropes, horses and locomotives. In many operations the loaded mine cars are hoisted to the surface, discharged and returned underground to repeat the cycle.

The manufacturers quoted here cater for a variety of load capacity and rail gauge requirements, making it possible to choose cars from the available range that can be used in an existing system. The car illustrated is a general purpose vehicle, nominally designed for use with a locomotive.

Operating information

There is a variety of available rail gauges and load capacities.

Source

Robert Hudson
 (Raletrux)
Division of Becorit
 (GN) Ltd
Oxclose Lane
Mansfield Woodhouse
Notts
NG19 3DF
UK

Telephone 0623 292061
Telex   55133

Other sources

MAN GHH
Bahnhofstrasse 66
Postfach 11 02 40
D-4200 Oberhausen 11
West Germany

Voest-Alpine AG
Postfach 2
A-4010
Linz
Austria

Huwood-Irwin Co.
PO Box 409
Irwin
PA 15642
USA

WHEELS AND AXLES

Using these wheel and axle sets (normally stocked as spares) enables custom-made chassis and bodies to be built up.

A wide range of sizes and types is available to suit particular applications. Wheels can be discs, spoked or double flanged, with diameters ranging from 152 to 914mm. A variety of wheel materials is available, the choice again depending on operating conditions.

Roller bearings are normally fitted for high speed or heavy loading applications.

A range of wagon couplings (not illustrated) is also available from the first manufacturer listed.

Source

Robert Hudson
 (Raletrux)
Division of Becorit
 (GN) Ltd
Oxclose Lane
Mansfield Woodhouse
Notts
NG19 3DF
UK

Telephone 0623 292061
Telex   55133

MAN GHH
Bahnhofstrasse 66
Postfach 11 02 40
D-4200 Oberhausen 11
West Germany

Voest-Alpine AG
Postfach 2
A-4010
Linz
Austria

Huwood-Irwin Co.
PO Box 409
Irwin
PA 15642
USA

Cast steel disc wheel

Cast steel machined for roller bearings

Solid disc wheel

Double flanged wheel

Spoked wheel and axles

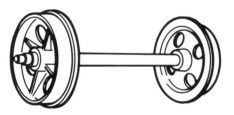

Disc wheels and axle with outside journals

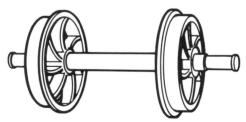

Spoked wheel and axles with outside journals

Disc wheels and axle with plain outside journals

TIPPLING MINE CARS

The body of a tippling mine car can be pivoted through 90 degrees from vertical in two directions. This allows the complete car load to be discharged to either side. The tippling procedure can be done manually or it can be integrated into a mechanical system.

Tippling cars are in general worldwide use, and are particularly useful in small haulages and small tonnage applications.

Operating information
A range of cars is available with different loading capacities. Modifications from the standard design are possible for individual applications.

Source
Robert Hudson   Telephone 0623 292061
(Raletrux)   Telex   55133
Division of Becorit
(GN) Ltd
Oxclose Lane
Mansfield Woodhouse
Notts
NG19 3DF
UK

Other sources
MAN GHH
Bahnhofstrasse 66
Postfach 11 02 40
D-4200 Oberhausen 11
West Germany

Voest-Alpine AG
Postfach 2
A-4010
Linz
Austria

Huwood-Irwin Co.
PO Box 409
Irwin
PA 15642
USA

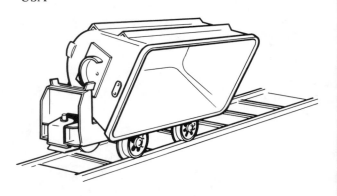

LOW PROFILE LOCOMOTIVE

Capable of operating in headings as low as one metre, this locomotive is battery powered, with the totally enclosed motor driving one axle through a worm reduction. Motor speed is controlled by a thyristor.

Braking is by a hand lever, and by plug braking with the speed handle reversed. Safety features include an isolator and a deadman's handle. Front and rear headlights are fitted.

Operating information
Two versions are available, with power rated at 1.6 or 2.2 kW.

Rail gauge	450 to 600mm
Minimum curve	4.3m radius
Tractive effort	Please refer to the graph
Hauling capacity	
1.6 kW version	9 tonnes at 9 kg per tonne resistance
2.2 kW version	11 tonnes at 9 kg per tonne resistance
Speed	
1.6 kW version	4.5 km per hour at 1 hour motor rate
2.2 kw version	8.0 km per hour at 1 hour motor rate
Battery	28 volt, lead acid, 203 Ampere hours at 5 hour discharge rate

Source
Pikrose & Co. Ltd   Telephone 061 370 1368
Wingrove & Rogers   Telex   667488
Division
Delta Road
Audenshaw
Manchester
M34 5HS
UK

Other sources
Clayton Equipment   Hunslet Group
Clayton Works   Hunslet Engine Works
Hatton   125 Jack Lane
Derbyshire   Leeds
DE6 5EB   LS10 1BT
UK   UK

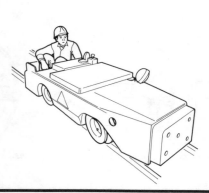

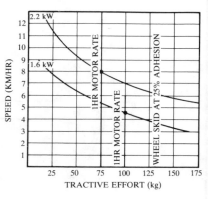

4-TONNE BATTERY LOCOMOTIVE

This small flame-proof battery locomotive is for use on
conventional track. Although there are smaller
locomotives on the market (around 2 tonnes), this size
is about the smallest that could be used for regular
production duties.

The locomotive is rubber tyred, and the resulting
considerable increase in traction allows operation on
gradients as steep as 1 in 10 (6 degrees).

Built-in electronics allow a functional readout to be
given, providing improved motor control, increased
reliability and the ability to diagnose faults.

A controller thyristor gives infinitely variable speed
control, coupled with dynamic plug braking. In all,
there are three braking systems:
— An electro-dynamic service brake
— A shoe-type parking brake on all wheels
— A deadman's brake, which is a spring applied disc on
   the cardan shaft.

Electric locomotives have advantages from a ventilation/
pollution aspect, but allowance must be made in costing
for a battery changing station. Although not included
here, a comparable range of diesel-powered locomotives
is also available.

Source
Becorit Ltd   Telephone 0602 302603
Hallam Fields Road   Telex   37526
Ilkeston
Derbyshire
DE7 4BS
UK

Other sources
Clayton Equipment
Clayton Works
Hatton
Derbyshire
DE6 5EB
UK

Hunslet Group
Hunslet Engine Works
125 Jack Lane
Leeds
LS10 1BT
UK

Operating information

Motor power	17.5 HP at 1180 rpm
Battery	114 volts (57 cells), 219 Ampere hours, OTM6
Speed	5 km per hour at 1 hour rating
Total tractive effort	2000 kg, maximum, at 506 adhesion
	823 kg at 1 hour motor rating
Weight	4 tonnes

BELT CONVEYOR DRIVE

This type of drive unit has been developed for and used exclusively in coal mining. However, it could clearly be used in any application which requires a fairly short, movable conveyor system.

The drivehead is a compact, two-drum geared tandem drive of medium power. It is used to power belt conveyors in development entries, and in production entries in partial extraction and longwall mining. It can also be used to convey minerals from the longwall on short length panels.

The unit is strong, but light in weight and easily manoeuverable. A loop take-up is available to provide adequate belt tension. Alternatively, a belt bank unit can be used to provide belt storage in addition to belt tension.

Operating information
Belt width	914mm
Belt speeds	1.6 or 2.3 m per minute (with 50 Hz supply)
	1.9 or 2.8 m per minute (with 60 Hz supply)
Driving drums	356mm diameter
Power	30 HP or 40 HP electric motor

Source
Dowty Meco Ltd   Telephone 0905 422291
Meco Works   Telex   338370
Worcester
WR2 5EG
UK

Other sources
Anderson Strathclyde plc
47 Broad Street
Glasgow
G40 2QW
UK

Babcock Mining Services Ltd
Bulman House
Regent Centre
Newcastle
NE4 4NG
UK

Roxon
Keskirkankaantie 9
SF-15860 Hallola
Finland

Note: there are very many conveyor manufacturers and suppliers, and most countries have their own internal sources.

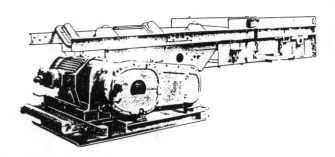

BELT CONVEYOR RETURN END

This is an anchored unit which provides a non-driven return for the belt.

The description 'return end' applies only to the belt, and not to the load to be transported. Because the return end receives the mineral load, it is strongly constructed, and fitted with impact-resistant idlers.

This return unit can be used with the drive already described, and the same manufacturer can also supply the other components (idlers and frame) needed to complete the conveyor. See also the next entry.

Operating information
Belt width	914mm or 1067mm versions available
Overall dimensions:	
— Length	2.905 m
— Height	549mm for 914mm width belt

Source
Dowty Meco Ltd   Telephone 0905 422291
Meco Works   Telex   338370
Worcester
WR2 5EG
UK

Other sources
Anderson Strathclyde plc
47 Broad Street
Glasgow
G40 2QW
UK

Babcock Mining Services Ltd
Bulman House
Regent Centre
Newcastle
NE4 4NG
UK

Roxon
Keskirkankaantie 9
SF-15860 Hallola
Finland

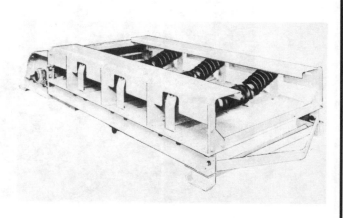

CONVEYOR BELT IDLERS

Although it is necessary to purchase the drives and return ends of conveyors, considerable amounts of money can be saved initially if the structure and belt idlers can be made locally. However, purchased rollers will probably have sealed-for-life bearings that need no lubrication, and provision has to be made for lubricating the home-made substitutes regularly.

The materials needed are angle iron, rods and pipes. Although a considerable amount of welding is necessary, the plans are simple to follow.

Operating information

Although the drawing gives specific dimensions, these will have to be scaled to suit the belt width.

The lower idlers which return the empty belt can simply be single straight rollers, since the belt does not need to be troughed to carry material.

Sources (manufactured idlers)

Dowty Meco Ltd
Meco Works
Worcester
WR2 5EG
UK

Telephone 0905 422291
Telex 338370

Anderson Strathclyde plc
47 Broad Street
Glasgow
G40 2QW
UK

Babcock Mining Services Ltd
Bulman House
Regent Centre
Newcastle
NE4 4NG
UK

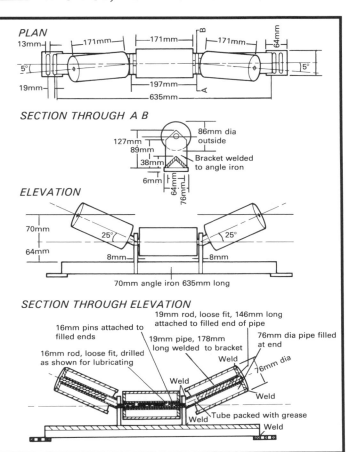

DIESEL/HYDRAULIC WINCH

This single drum winch has independent power supplied by a Lister/Petter air-cooled diesel engine. The built-in hydraulic motor provides a fail-safe braking system. The unit is mounted on skids, which facilitate siting and transport.

This type of winch can be used for a variety of haulage applications, and is especially suitable in areas remote from power supplies, or in places where there is familiarity with the maintenance of diesel prime movers.

Operating information

Safe working load   5 tonnes (mid layer)
Line speed   0 to 10m per minute
Rope capacity   200m of 19mm diameter wire
                           rope

Hydraulic pressure   300 psi
Control   Forward/neutral/reverse

Source

Pikrose & Co. Ltd
Delta Works
Delta Road
Audenshaw
Manchester
UK

Telephone 061 370 1368
Telex 667488

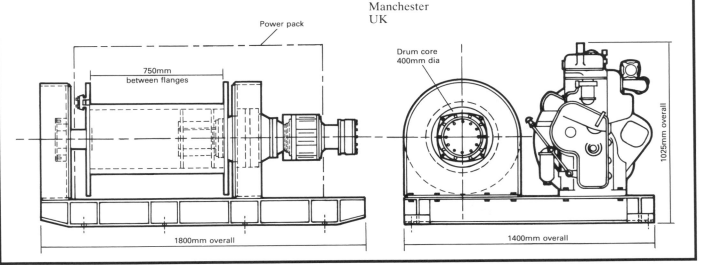

AIR WINCH

The air winch is simple and robust, readily portable, and operates equally well horizontally and vertically. It can be bolted to a floor or foundation, or it can be clamped to a column or bar (as used for mounting rock drills). It can also be bolted to a wall or the leg of a crane.

Power is provided by a rotary piston air motor, giving maximum torque at starting. The winch is particularly suitable for hoisting and haulage work. A brake is fitted which can also be used to control lowering speed, and there is also a forward-stop-reverse control lever.

Operating information

Power	Compressed air supply required
Performance	
— Air pressure	Figures below relate to 5.6 kg per cm^2
— Mid coil	1079 kg at rope speed of 12.2 m per minute
	716 kg at rope speed of 42.6 m per minute
— Top coil	862 kg at rope speed of 15.3 m per minute
	571 kg at rope speed of 53.3 m per minute
Drum size	152mm diameter, 203mm wide, 83mm flanges
Capacity	
— Mid coil	44 m of 11mm diameter rope
	52 m of 10mm diameter rope
	78 m of 8mm diameter rope
— Top coil	91 m of 11mm diameter rope
	125 m of 10mm diameter rope
	183 m of 8mm diameter rope

Source
Enquiries to registered and sales office:
Holman-Climax   Telephone 44-8648/44-9156
  Manufacturing Ltd
Dolphin Court
2nd Floor
7A Middleton Street
Calcutta-700 071
India
(Holman-Climax is a subsidiary of Compair International Ltd, Slough, UK)

Other sources
Atlas Copco MCT
S-104 84
Stockholm
Sweden

Ingersol Rand Co.
  — IMPCO
150 Burke Street
Nashua
NH 03061
USA

Joy Manufacturing Co.
  UK Ltd
Peel Park Place
College Milton
East Kilbride
Glasgow
G74 5LN
UK

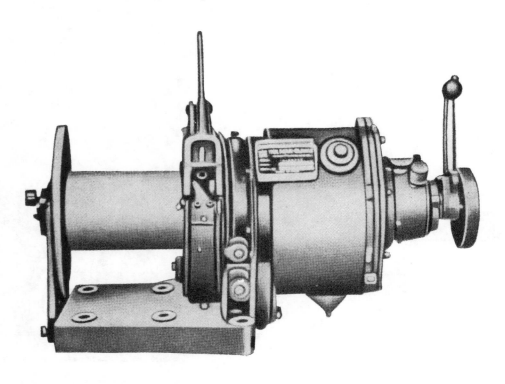

3-TONNE ELECTRIC WINCH

A standard electric motor provides the drive for this single drum winch, which is suitable for a variety of haulage operations. Many other prime mover variants are possible, including compressed air, diesel, diesel/hydraulic, and (for gassy mines) flameproof electric motors.

The unit has a manual braking system, and a fail-safe electric system is available at cost.

Attachments allow the winch to be used for more specific applications. For example, the addition of a surge wheel makes the unit suitable for endless haulages and with the addition of fail-safe braking, the winch could be used as a service hoist in a small operation.

Operating information

Drum capacity	168mm of 16mm diameter wire rope, which is wound in 7 layers
Mean rope speed	0.25 metres per second
Mean rope pull	3 tonnes
Brake	Manual

Ropes are not included.

Source

Pikrose & Co. Ltd
Delta Works
Delta Road
Audenshaw
Manchester
M34 5HS
UK

Telephone 061 370 1368
Telex   667488

Other sources

Markham and Co. Ltd
Broad Oaks Works
Chesterfield
Derbyshire
S41 0DS
UK

Joy Manufacturing Co. UK Ltd
Peel Park Place
College Milton
East Kilbride
Glasgow
UK

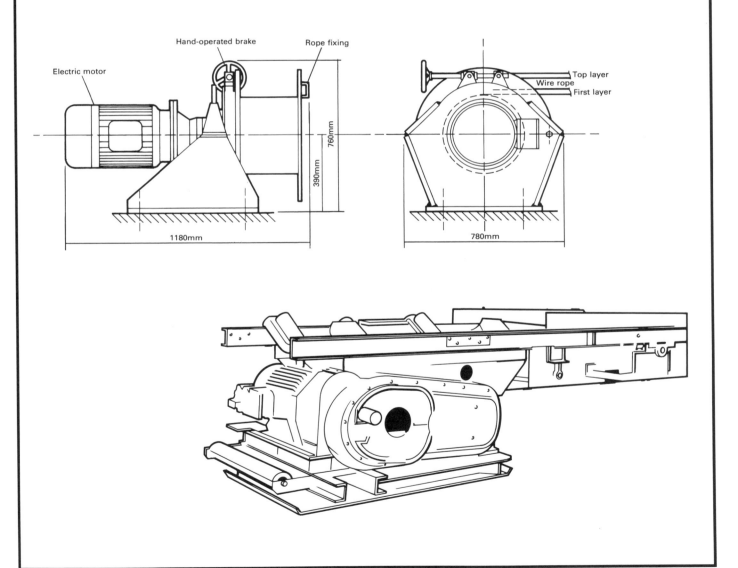

HOIST FOR SMALL SHAFTS

This automatic hoist is adequate for small, shallow vertical shafts in small tonnage mining operations. The steel frame incorporates the winding and tipping gear, and a ladder which extends 49 feet (14.9 metres) down the shaft. Additional 7 feet (2.1 metres) ladder lengths are available.

The equipment is lightweight and portable, but durable and completely self contained. Thus it can easily be repositioned when required.

Operating information

The hoist is supplied complete with motor and cable.

Suitable for one man operation, the unit hauls a 12 gallon drum, which is tipped over the top of the frame (as illustrated) for direct loading into a truck or other arrangement.

Source

Miner's Den
464 Whitehorse Road
Mitcham
3132 Victoria
Australia

Telephone 03 873 1244
Telex   AA36521
            FROMEL

Or fabricate locally.

HEADFRAME AND WINDER EQUIPMENT

Qualter Hall is a company which manufactures equipment and undertakes turnkey installation contracts for mineshaft equipment, headframes and winders (including skips and mancages) and underground and surface distribution systems.

Senior staff of this company have written a number of technical papers. Important among these are:

Newsum, J. Mineshaft Hoists, *Colliery Guardian*, February 1980.

Newsum, J. Headgear Overwind Catchgear, *Colliery Guardian*, July 1982.

Newsum, J. Mineshaft Installations, *The Mining Engineer*, August 1988. This article deals with larger skip winding plants, but contains much general practical advice and technical data.

Source

Qualter Hall & Co. Ltd
PO Box 8
Johnson Street
Barnsley
South Yorkshire
S75 2BY
UK

Telephone 0226 205761
Telex   54697

MINE WINDERS AND HAULAGE WINCHES

Generally speaking, these pieces of equipment are robust, continuously-rated machines. Size must be chosen to suit particular mine requirements.

The photograph shows a 55 kW, 1000mm diameter winder for incline haulage (men) as supplied by this manufacturer to a tin mine in the United Kingdom.

The unit illustrated is supplied by GEC who make a range of mine winders, haulages and braking systems. Their range of winding equipment uses proven techniques, with conventional drum winders and multi-rope friction machines.

Other equipment available from this particular source includes complete mineral processing installations or individual processing machines for crushing, sizing, attrition, comminution or beneficiation.

Operating information

Much of this company's output (but not all) is in heavier machines which will not interest small-scale miners. The following table lists their standard range of electro/hydraulic winches, for which mining power packs are available.

SIZE	ROPE PULL		ROPE SPEED		ROPE DIA.		ROPE CAP.		DRUM DIA.		HP
	tonnes	kgs	ft/min	m/min	inch	mm	ft	mtrs	ft	mm	(kw)
3.5/950	3.50	3500	300	91.4	0.63	16	3280	1000	3.117	950	100 (75)
5.0/1150	5.00	5000	300	91.4	0.75	19	3280	1000	3.773	1150	150 (110)
6.5/1350	6.50	6500	300	91.4	0.87	22	3280	1000	4.429	1350	200 (150)
8.0/1500	8.00	8000	300	91.4	0.94	24	3280	1000	4.921	1500	225 (168)
10.0/1650	10.00	10000	300	91.4	1.06	27	3280	1000	5.413	1650	300 (220)
12.0/1800	12.00	12000	300	91.4	1.18	30	3280	1000	5.905	1800	350 (264)
15.0/2000	15.00	15000	300	91.4	1.30	33	3280	1000	6.562	2000	450 (336)
18.0/2200	18.00	18000	300	91.4	1.46	37	3280	1000	7.218	2200	500 (370)

Source

GEC Mechanical
  Handling Ltd
Mechanical Process
  Division
Cambridge Road
Whetstone
Leicester
LE8 3LH
UK
(Formerly known for over 100 years as Fraser & Chalmers.)

Telephone 0533 863434
Telex   347344 or 347539

Other sources

Rexnord Inc. Process
  Machinery
3500 First Wisconsin
  Centre
777 East Wisconsin
  Avenue
Milwaukee
Wisconsin
53202
USA

MINE WINDER

Illustrated is a friction winder which was manufactured by NEI Peebles for a coal mine in Britain. Although this example is rated at 855 kW, and the company is capable of making winders with ten times this rating, they can supply winders with motors rated as low as 37 kW.

The company will undertake complete turnkey projects, with follow-up inspections and service. They will also carry out rehabilitation or modernization of existing AC or DC mine winder systems.

Note: there are two basic parts to winders:
(a) mechanical and (b) electrical plus control. Not all companies manufacture both, so it is common for the complete machine to incorporate (for example) Markham (mechanical) and NEI (electrical) parts.

Operating information

Range available	From 37 kW up to 8000 kW
Drive	Electric motors, either AC or DC, direct coupled or geared
Controls	Microprocessor, with safety circuits

Source

NEI Peebles Ltd
Peebles Electrical
  Machines
581 Tyburn Road
Erdington
Birmingham
B24 9RX
UK

Telephone 021 384 6644
Telex   339177

Other sources

ASEA Metallurgy
(now ABB)
Avd Fas, S-72183
Vasteras
Sweden

Rexnord Inc. Process
  Machinery
3500 First Wisconsin
  Centre
777 East Wisconsin Ave
Milwaukee
Wisconsin 53202
USA

MAN GHH
Bahnhofstrasse 66
Postfach 11 02 40
D-4200 Oberhausen 11
West Germany

Markham & Co. Ltd
Broad Oak Works
Chesterfield
Derbyshire
S41 0DS
UK

CAGES AND SKIPS

The supplier of the cages and skips illustrated offers a large range of haulage and minecar equipment. Cages and skips are made in many sizes, the smaller of which are illustrated here.

The upper picture is an aluminium alloy two-deck cage fitted with safety gripper gear. In this cage the centre deck can be hinged upwards, so that long lengths of material can be carried.

In the lower illustration an ore skip of the overturning type is shown. This is also constructed of aluminium alloy and has a capacity of 1.85 cubic metres (65 cubic feet).

Source

Known by the manufacturer's name of Allens of Tipton, these units are now produced at:

Jenkins of Retford Ltd
Thrumpton Lane
Retford
Nottinghamshire
DN22 7AN
UK

Telephone 0777 706777
Telex   56122

Other sources

MAN GHH
Bahnhofstrasse 66
Postfach 11 02 40
D-4200 Oberhausen 11
West Germany

Markham & Co. Ltd
Broad Oak Works
Chesterfield
Derbyshire
S41 0DS
UK

Rexnord Inc. Process Machinery
3500 First Wisconsin Centre
777 East Wisconsin Avenue
Milwaukee
Wisconsin 53202
USA

Qualter Hall & Co. Ltd
PO Box 8
Johnson Street
Barnsley
South Yorkshire
S75 2BY
UK

ROPES FOR HAULAGE AND WINDING

Many different kinds of wire ropes are used in surface and underground mining.

Each cross section style has its own characteristics. For example, locked coil ropes are specially designed to be non-rotating under normal conditions and they can withstand high radial and compressive forces (making them ideal for winders and hoists).

Operating information

Hoist ropes (see diagram)		Size range (mm diam.)	Breaking load range (minimum in tonnes)
Hoist	1	13-64	9-249
	2	13-70	10-396
	3	8-64	4-250
	4	16-65	22-362
	5	35-76	45-218
	6	29-51	43-133
Haulage	A	8-38	3- 92
	B	13-41	9-105
	C	13-16	10- 16
	D	26-60	40-225

Source
British Ropes Ltd
Carr Hill
Doncaster
DN4 8DG
UK

Telephone 0302 344010
Telex   547981

Other sources
Bruntons (Musselburgh) plc
Inveresk Road
Musselburgh
East Lothian
EH21 7UG
UK

Usha Martin Black Ltd
14 Princes Street
Calcutta 700 072
India

Cross sections of hoist ropes

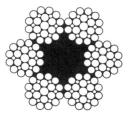

1 Round strand

2 Triangular strand

3 Multistrand, non-rotating

4 Locked coil

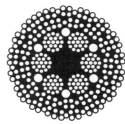

5 Superflex multi-flat stranded (non-rotating)

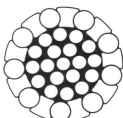

6 Guide and rubbing ropes

Cross sections of haulage ropes

A Round strand

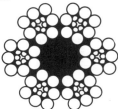

B Triangular strand

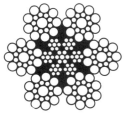

C Coal cutter and slusher ropes

D Cable bent ropes

6. Safety equipment

Mining, by its very nature, has always been a hazardous business. Thus the extent and effect of the hazards have never borne any relation to the size of the operation, and a worker on a small-scale mining enterprise faces at least as much danger as one on a large operation, where more funds may be available for sophisticated safety equipment and extensive preventive measures. The importance of proper safety precautions is perhaps even greater in a small mine, because of the disproportionate effect of a mishap on specialist manpower and/or production.

Investment in safety measures and equipment should never be regarded merely in the light of necessary compliance with statutory regulations. It should always be seen as a vital adjunct to improved productivity, through higher workforce morale, fewer stoppages and shutdowns, improved workforce/management relationships, and resultant greater efficiency. Thus safety-consciousness can be a positive move towards higher profitability.

Safety legislation has in recent years tended to distribute responsibility more widely than hitherto, giving workers at every level of authority their due degree of the burden. Any such stimulation of the awareness of hazards in the workplace must have a salutary effect, but the weight of responsibility still lies with management. Hazards may be identified or anticipated, and equipment may be brought in to circumvent the dangers, but however efficient or sophisticated the equipment, it is only as good as the person who uses it. Top priority must be given to training and regular practice in its use. Furthermore, management should establish orderly shutdown techniques and evacuation procedures, and arrange training in the practice of first aid and fire-fighting, and in the safe movement of casualties.

Every member of the workforce, whether employed above or below ground, should be conversant with the site safety procedures as they may affect himself, and management should ensure that regular drills are held to ensure that a dangerous situation will be dealt with calmly and efficiently.

The dangers inherent in mining lie at the workplace, in the risks attached to extracting and handling the ore and other materials, and in possible misuse or malfunctions of equipment. There is also, of course, the danger of disturbance to the environment from the mining operation.

Several companies have developed instruments aimed at identifying hazards at the workplace, and the miner's oldest enemy, unsafe air, can now be fought by a range of sophisticated devices which will measure the proportion of oxygen in the air, as well as monitoring levels of methane and other toxic gases. Examples and descriptions of such equipment, all readily portable, can be found on the following pages.

In the event of vital work being necessary in a contaminated environment, such as the recovery of a casualty, breathing apparatus with a working duration of up to four hours is now available. Sets of such apparatus should be held both underground and at collar level, and should be tested regularly.

Personal protection such as safety helmets and steel-capped boots should be part of the normal dress of every mining worker, but the use of ear protectors should also be encouraged where noise levels are excessive or prolonged. Comfortable safety goggles should also be part of each man's personal equipment, and an adequate supply of respirators should also be provided for use in excessively dusty environments.

Dust is an occupational hazard for the miner, and can be the cause of long-term illness if suitable precautions are not taken.

Fortunately equipment is available to minimize the menace, and more than one company specializes in dust-proof materials handling by means of tubular conveyors. Retractable loading spouts, which reduce the dust hazard when transferring dry bulk material, also make a considerable contribution to the comfort of the workplace. Any doubt over the acceptability of ambient dust levels may be resolved by the use of hand-held dust monitoring instruments, which can be calibrated to respond to various types of dust and fibre.

While still at the workplace, a large proportion of accidents are known to occur through the most easily corrected cause, that of bad housekeeping.

Whilst the most effective remedy is good discipline, a useful aid is available in the shape of vacuum systems, with take-off points positioned at areas of hazardous or maximum spillage. In the interests of minimizing the fire hazard, waste timber and other combustible rubbish should never be allowed to accumulate. The storage of flammable materials underground is of course strictly controlled by mining regulations.

In view of the inevitably serious effects of a fire below ground, the importance of suitable fire detection and protection systems cannot be over-emphasized. Fire-fighting equipment available ranges from complex ring-main installations to portable appliances, and generous provision of equipment to cope with gas, oil, and electrically-seated fires is strongly recommended.

Good communications are vital to efficient and safe mine working, especially underground, and the installation of adequate signalling systems which can be understood by every member of the workforce is imperative. Some modern systems, by the use of safety interlocks, pre-empt the operator's decision-making process, saving vital time by putting a series of events in train immediately on the detection of a potential hazard. The detection of an anomaly in the haulage rope tension pattern, for example, will result in the halting of the winding operation to enable the problem to be investigated.

Arrestor systems for conveyances which have become hazardous, either through excessive speed or through a braking malfunction, have been produced by a number of shaft equipment manufacturers. One which combines effectiveness and safe deceleration with minimum maintenance depends simply on the conversion of the kinetic and potential energy of the conveyance into strain energy and heat generated by the deformation of flat strips of metal.

Awareness of the importance of safety in mines is nothing new. The basis of all national mining regulations has always been sound, safe and sensible working practices. In all such regulations will be found safety stipulations based on firm empirical foundations for such things as design loading factors for lifting devices, the storage and handling of explosives, the installation of toeboards and handrails and a myriad other matters for the protection of the workforce.

It could be argued that mishaps in small mines are less excusable than in larger operations. The structure of responsibility is less diffuse, lines of communication are shorter, the equipment is on a more human scale, and the whole operation is more manageable. But safety in small mines, as in any other hazardous occupation, can only be achieved through the installation, use and maintenance of appropriate equipment, proper training, suitable discipline and, above all, the right attitude at every level.

Alan Bartley

CAP LAMPS

An electric cap lamp which can be fitted on to standard helmet brackets is available from the company listed below (which also stocks the helmets). The rechargeable nickel cadmium battery is housed in a separate shockproof case, provided with a lockable cover. The battery case is carried at waist level, by attaching it to a belt with the loop provided.

Operating information
The electrolyte is non-acidic and there is protection against tipping. Batteries can be recharged about 800 to 1,000 times. The bulb has two filaments, allowing two beam intensity settings (high and low), and the beam is also tiltable. A magnetic lock and screws prevent accidental opening in gassy mine atmospheres.

Bulb life	400 to 500 hours
Battery dimensions	127mm × 58mm × 180mm
Weight	2.4Kg

Source
Miners Inc.   Telephone 208 628 3247
PO Box 1301   Telex   150030
Riggins
Idaho 83549
USA

A cap lamp with integrated personal methane alarm is available from:

Oldham Batteries Ltd   Telephone 061 336 2431
Denton   Telex   668321
Manchester
M34 3AT
UK

Also available, a lightweight miners' cap lamp, watertight and featuring miniature tungsten halogen bulbs. The battery, contoured to fit the body, is half the weight of conventional batteries, and has a life expectancy of 1,000 recharge cycles. Available from:

Levitt-Safety Ltd   Telephone (416) 425 8700
33 Laird Drive   Telex   06-22286
Toronto
Ontario
M4G 3S9
Canada

GAS DETECTOR — OXYGEN DEFICIENCY

In addition to measuring the percentage volume of oxygen present, this hand-held detector will give an audible warning if the oxygen level should fall below a preset amount.

The instrument is safe for use in underground mines and potentially explosive atmospheres.

Two versions are available, one reading the percentage of oxygen present by volume and the other calibrated to show the partial pressure.

Operating information
Battery life	250 hours continuous operation from full charge
Measuring range	Zero to 25 per cent of oxygen by volume
Alarm range	14 to 23 per cent of oxygen by volume
Accuracy	Plus or minus 0.5 per cent of oxygen
Dimensions	188 × 78 × 33mm
Weight	700g
Environment:	
— Humidity	Zero to 100 per cent relative humidity
— Temperature	Zero to 40 degrees Celsius

Source
Sieger Ltd   Telephone 0623 513222
UK Sales Office   Telex   377143
Fulwood Close
Fulwood Industrial Estate
Sutton-in-Ashfield
Notts
NG17 2JZ
UK

Pocket oxygenometers are also available from:

Gas Detection Division   Telex 133 227
BP 962
62033 Arras Cedex
France

Hand-held portable oxygen monitoring instruments and toximeters are available from:

Dragerwerk AB Lubeck   Telephone 494 518820
Postfach 1339
D-2400 Lubeck
West Germany

GAS DETECTOR — CARBON MONOXIDE

This hand-held unit is robust and intrinsically safe for use in potentially explosive atmospheres. It is powered from an internal battery, the charge of which is conserved by means of an automatic cut-off switch.

A digital readout is used to display gas levels, with low battery and over-range indications. Performance is most stable in the range up to 30ppm, but levels up to 200ppm can be indicated.

A filter is fitted to remove unsaturated hydrocarbons before they can reach the electrochemical sensor cell.

Operating information

Battery	Nickel cadmium cell, giving approximately 2,000 readings from a full charge
Readout type	3 digits, LCD
Range	Zero to 200ppm total range, with stability best over the range zero to 30ppm of carbon monoxide
Weight	650g

Source

Sieger Ltd   Telephone 0623 513222
UK Sales Office   Telex   377143
Fulwood Close
Fulwood Industrial Estate
Sutton-in-Ashfield
Notts
NG17 2JZ
UK

GAS DETECTOR — METHANE

This hand-held unit is robust and intrinsically safe for use in potentially explosive atmospheres. It is powered from an internal battery, the charge of which is conserved by means of an automatic cut-off switch.

The digital readout gives the percentage of methane, and can also indicate low battery, fault, or a gas level which exceeds the instrument's range.

Calibration is simple, using a side-mounted adjustable potentiometer.

Operating information

Battery life	Better than 200 tests of 40 seconds duration
Readout type	3 digits, LCD
Range	Zero to 5 per cent of methane by volume
Accuracy	Plus or minus 0.1 per cent by volume, or 8 per cent of the reading, whichever is the greater
Dimensions	190 × 80 × 45mm
Weight	650g

Source

Sieger Ltd   Telephone 0623 513222
(as above)

A pocket methanometer similar to that illustrated here is available from:

Gas Detection Division   Telex 133 227
BP 962
62033 Arras Cedex
France

ARRESTOR GEAR FOR LARGE MOVING MASSES

Using a novel method of energy absorption, this is an invention which has been installed in a number of mines throughout the world to arrest the movement of cages when overtravelling. For drum winders the device can be installed at the bottom of the shaft, with conventional safety gear used at the top. With friction winders these arrestors can be fitted at both ends of the wind to give complete protection in both directions of travel.

Since the device can bring any large mass safely to rest, it also has application for intercepting and stopping runaway mine cars and locomotive-hauled trains.

The principle of the arrestor is a corrosion resistant mild steel strip which is threaded through a set of stainless steel rollers in such a way that it must be deformed. One end of the strip is securely anchored, so that the strip is pulled through the rollers when the arrestor plate or the beam to which it is attached is struck by the moving mass. The deformation pattern thus travels along the strip, absorbing the kinetic and potential energy of the mass, dissipating the heat generated over a large area of steel strip and bringing the mass to a safe stop with linear retardation and no risk of catastrophic deceleration. Results are accurately predictable and unaffected by oil, grease or water.

The manufacturer will calculate the cross section and length of steel strip needed, and the size and layout pattern of the rollers. For mine cage applications two or more strips are suspended from the shaft steelwork, and the roller boxes for each are mounted on the arrestor beam.

Known as the Selda Arrestor, it can be installed easily in existing locations. All roller boxes are sealed against the ingress of dirt and grit, and all parts are either stainless steel or corrosion protected. The manufacturer calibrates each device and preforms the strips before shipping.

Operating Information
Roller boxes are made in various sizes, designed to cover a range of forces from 30KN up to 350KN each.

Source
Fairport Engineering Ltd   Telephone 0257 483311
Market Street   Telex   67425
Adlington
Lancashire
PR7 4EZ
UK

Details of cage arrestor units are available from Davy McKee and Qualter Hall:

Davy McKee   Telephone 0642 602221
  (Stockton) Ltd   Telex   587151
Ashmore House
Stockton-on-Tees
Cleveland
TS18 3RE
UK

Qualter Hall & Co. Ltd   Telephone 0226 205761
PO Box 8   Telex   54697
Johnson Street
Barnsley
South Yorkshire
S75 1BY
UK

Other equipment
Shaft safety systems, including winder monitoring, rope slip and slack rope protection; also intrinsically safe shaft signals and interlocks. Further details and equipment from:

Transmitton Ltd   Telephone 0530 415941
Smisby Road   Telex   342284
Ashby-de-la-Zouch
Leicestershire
LE6 5UG
UK

Direction of travel

Top end of strip anchored to shaft steelwork

Steel strip

Roller box attached securely to arrestor beam

Strip deformed over stainless steel rollers

Sufficient length of strip to take up stopping distance

*Selda principle
(mine cage application)*

EQUIPMENT FOR PERSONAL PROTECTION

The importance of personal protective gear has already been emphasized. We have received information on a comprehensive range of personal safety equipment and clothing, including:
- ☐ Safety boots (with steel toecaps)
- ☐ Ear protectors
- ☐ Safety helmets
- ☐ Transparent vizors and browguards
- ☐ Safety goggles
- ☐ Steel face masks for welding
- ☐ Respirators
- ☐ Body harnesses and safety slings

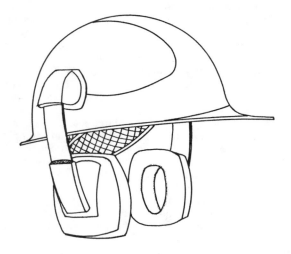

Source

Seguridad Eléctrica S.A.   Telephone 22700
Maipu 265
5000 Cordoba
Argentina

Hearing protectors, including earplugs and earmuff types, together with education in use-motivation methods manufactured by:

Cabot Safety Ltd   Telephone 06025 878320
E.A.R. Division   Telex   665303
First Avenue
Poynton
Cheshire SK12 1YT
UK

Breathing apparatus for escape from hazardous environments, including automatic oxygen-breathing set for use by mine rescue teams, combining low weight with 4-hour working duration. Filter self-rescuer, giving protection against carbon dioxide and combustion gases for up to 3 hours, with good cooling of the inhaled gases, and low breathing resistance. Available from:

Dragerwerk AB Lubeck   Telephone 494 518820
Postfach 1339   Telex   26807
D-2400 Lubeck 1
West Germany

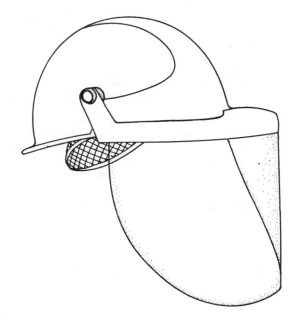

Rescue/escape breathing apparatus for use in toxic or oxygen-deficient atmospheres, with closed-circuit device using potassium superoxide to produce oxygen from exhaled CO_2 and moisture. Available from:

MSA (Britain) Ltd   Telephone 0236 24966
East Shawhead   Telex   778396
Coatbridge
ML5 4TD
UK

7. General purpose equipment

This section covers equipment and tools which can be used in more than one section of a mining operation. For ease of reference they are categorized under the headings of Tools, Prime movers and generators, Air compressors and Pumps.

Tools

The type, size and diversity of hand and machine tools needed for maintenance purposes to ensure continuous and continuing mining and process operations depends on a number of factors, the most important being the location of the mine. For example, is there any industrial infrastructure close by?

The provision of a workshop facility is essential in any mine, and if potential owners/operators are in any doubt as to requirements, expert advice should be sought. The list of tools in this section is not comprehensive and is only an indication of the basic requirements.

Prime movers and generators

Under normal circumstances it is best if prime movers (e.g. electric motors, internal combustion engines) are purchased as a unit with the machine being driven to ensure they are matched to the design duty. The most common type of electric motor used throughout mines is the 'squirrel cage' type which is very robust and of simple construction.

If there are a number of electric motors of the same frame size and horse power it is prudent to keep at least one spare motor. Large electric motors are usually for specific applications and if purchased separately from the driven unit it is essential that potential suppliers are given a comprehensive duty specification before buying.

Small portable generators of the sizes listed are of great use, particularly on mines with a limited electrical distribution system.

Air compressors

Compressed air in mining applications is mainly used to provide power for tools and equipment which operate with an air pressure of about 0.5 to 0.6 MPa (75 to 90 psi), but to allow for pressure losses in pipes/hoses, compressors should be rated at between 0.7 to 0.8 MPa (100 to 110 psi). Compressor volume is usually expressed in terms of free air delivered (FAD), that is air at ambient conditions, before compression. However, the correct definition of capacity is under Normal Conditions, i.e. 0.1 MPa or 1 bar at 0°C. In America, capacity is often expressed under Standard Conditions, which is 0.1 MPa at 15.5°C. For these reasons it is important to state capacity as FAD, N or S.

Vane-type compressors (rotary) are usually preferred to cylinder types (reciprocating), and packaged units including driver, compressor inter/after coolers, water separator and filters and control panel, all in a sound insulating enclosure, are widely available in most countries. Examples of the types available are shown on the following pages.

Pumps

There are many types of pump, each particularly suitable for certain applications. Pump types include centrifugal or rotodynamic, rotary positive displacement such as gear, single or twin helical screw, lobe, peristaltic and disc, also reciprocating positive displacement types such as piston, plunger or diaphragm. This introduction deals only with centrifugal pumps which are, despite their limitations, the most commonly used within the mining industry.

There are many types of centrifugal pump and, as a first division, they can be separated into those designed to handle clear liquids such as water and those that can handle solids in suspension.

In mining applications, clear liquid pumps are mostly used for water. Efficiency varies with pump size, smaller pumps being less efficient than large ones, pumps with a discharge nozzle size smaller than 50mm having efficiencies in the region of 50 per cent and larger pumps with efficiencies of around 80 per cent. Pumps should be selected for operation at BEP (Best Efficiency Point). For water, the usual material for both casings and impellers is cast iron, but bronze impellers are frequently used. The most practical seal arrangement for mining applications are packed glands. Various drivers can be used; electric motors are the most convenient, but internal combustion drivers can be used when electric power is not available.

It is recommended that pumps are purchased as a unit complete with driver, mounted on a baseplate. In all cases, manufacturers' pump curves should be carefully studied to ensure a choice that will fit the full range of operating conditions required. Before selection, it is advisable to draw a system curve for the operating conditions to obtain the best match with the pump operating curves. Centrifugal pumps absorb more power at the lower end of the curve, and motors should therefore have sufficient power under conditions of lowest head and high volume.

Centrifugal pumps for use with slurries, that is liquids containing solids in suspension, require careful selection. Slurry pumps can handle up to 60 per cent solids by weight; maximum particle size depends on pump size and construction. Pump curves usually include the maximum particle size that can be handled. A good subdivision would include three main categories:
— Drainage pumps
— Sand pumps
— Gravel/dredge pumps.

Drainage pumps are usually submersible, with an integral motor cooled by the liquid pumped, but the pumps can run for limited periods 'on snore', that is dry or partially dry, without overheating. These pumps are not true slurry pumps, being designed to pump essentially dirty water, and should not be used with solids concentrations in excess of 30 per cent by weight.

Sand pumps are intended for use with slurries having an average particle size about 100 microns, but can be used with larger particles providing the material is not too sharp. Hard metal-lined sand pumps can handle particles up to 4mm dia., but where larger particles are predominant, gravel pumps are preferred. For fairly fine and not particularly sharp particles, rubber linings and rubber impellers give longer life, but where large particles with sharp edges and high hardness are present, hard metal liners and impellers are preferred.

Selection of size and type for slurry pumps is very important, as use away from BEP creates additional turbulence which results in higher wear. Slurry pump discharges should never be throttled, and variations in duty should be achieved by speed variation. Where duty variations are infrequent, a pulley drive gives adequate flexibility, as pulley sizes can be changed. Otherwise variable speed drives are recommended.

The efficiency of slurry pumps is lower than the efficiency of water pumps, a good average being 60 per cent. Small pumps wear quite rapidly, and the efficiency of pumps with a discharge nozzle of 40mm can be as low as 30 per cent. Usually it is preferred to avoid the use of slurry pumps with a discharge nozzle below 50mm. In general, good slurry pumps should be of solid construction with massive bearings to accommodate heavy impellers and eccentric loading due to possible uneven solids distribution.

Some examples of the types available are shown on the following pages.

Ernest Hogg and Stach Odrowaz-Pieniazek

GENERAL PURPOSE HAND TOOLS

We have received information on a small selection of hand tools from the supplier listed below. The tools are:

Geologist's rock hammer
The Estwing geologist's rock hammer, used for sample collecting during surveys, has a pointed pick at one end and a 19mm hammer face at the other. The head is forged in one piece, and the handle is made from hardwood, with a leather grip. The weight is 0.6 Kg and the handle is 330mm long. Price is approximately $24 US.

Hand pick
This is an example of a hand held pick/matlock for breaking or trenching and costs about $20 US. The dimensions of this example are:
— Handle length   914mm (hardwood)
— Head span   457mm from pick to matlock
— Matlock width   50mm
— Weight   2.36 Kg

Coal pick
This pick is manufactured by Warwood Forged Steels and costs around $27 US. It is designed specially for coal use, with a double-ended pointed pick taper fitted to a hardwood handle. Its dimensions are:
— Pick span   432mm, point to point
— Length   863mm
— Weight   2.27 Kg

Source
For these, and similar products:

Miners	Telephone 208 628 3247
PO Box 1301	Telex 150030
Riggins	
Idaho 83549	
USA	

LEVER LIFTING TOOL

A simple, hand-operated tool for lifting, pulling, tensioning and securing. The device is useful in confined spaces.

The body is made of aluminium alloy, for portability. Safety catches are provided on all hooks and, to facilitate rapid use, a wind-and-pull-through mode can be selected.

Operating information

Safe working load	1.5 tonnes
Weight	13.6 Kg, with 3m of chain
Velocity ratio	54.5:1
Safety factor	5:1
Handle length	419mm
Body length	413mm
Hook clearance	29mm

Source

Davy Morris Ltd	Telephone 0509 610061
PO Box 7	Telex 34408
North Road	
Loughborough	
Leicestershire	
LE11 1RL	
UK	

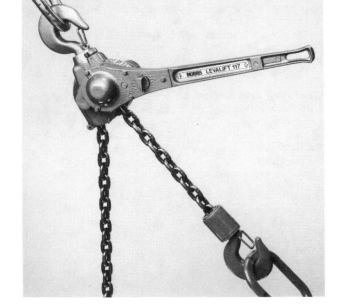

HAND CHAIN HOIST

A light and strong hand-operated chain hoist designed particularly for mining environments. Safety clips on the top and bottom hooks prevent accidental slipping.

The load is raised by a hand chain, which is separate from the lifting chain attached to the load.

Operating information

Safe working load	1 or 2 tonne hoists are available — details below are for the 1 tonne hoist
Effort required	26 Kg to raise 1 tonne
Weight	13 Kg, with 3m of chain
Velocity ratio	45:1
Hook admittance	32mm

Source

Davy Morris Ltd
PO Box 7
North Road
Loughborough
Leicestershire
LE11 1RL
UK

Telephone 0509 610061
Telex   34408

MONORAIL-MOUNTED HAND CHAIN HOIST

This chain hoist has a travelling block designed for mounting on a monorail constructed from suspended I or inverted T section girders.

The robust and rigid unit is sealed against dirt and has lifelong lubrication. Guides ensure smooth chain travel.

The travelling block has a self sustaining brake.

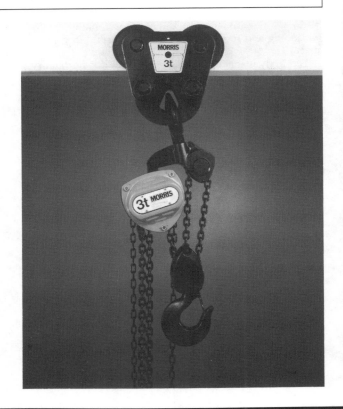

Operating information

Safe working load	3 tonnes
Effort required	31 Kg, to raise 3 tonnes
Weight	27 Kg, including 3m of chain
Safety factor	3:1
Velocity ratio	135:1 (3 falls of chain)
Minimum track radius	1.8m

Source

Davy Morris Ltd
PO Box 7
North Road
Loughborough
Leicestershire
LE11 1RL
UK

Telephone 0509 610061
Telex   34408

GAS WELDING AND CUTTING EQUIPMENT

A combined welding and cutting outfit for welding up to
8mm and cutting up to 50mm, complete in a carrying
case. Regulators and hoses are usually purchased
separately, but are included here as a complete
package.

Source
Buck & Hickman Ltd   Telephone 0742 766660
Bank House
100 Queen Street
Sheffield
S1 2DW
UK

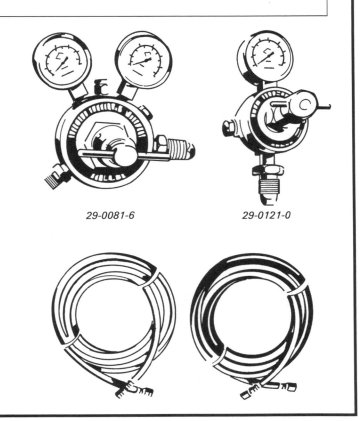

29-0081-6   29-0121-0

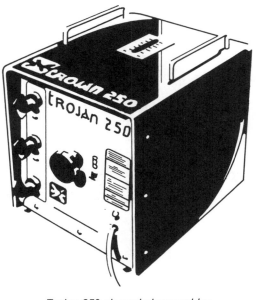

ELECTRIC WELDING EQUIPMENT

Electric welding sets are available in a wide range of
capacities, and any special requirement should be
included in the enquiry specification to potential
suppliers. The one illustrated would be adequate for
most 'run of the mill' works on a mine.

Source
Buck & Hickman Ltd   Telephone 0742 766660
Bank House
100 Queen Street
Sheffield
S1 2DW
UK

Operating information

Electrode sizes	1.6mm to 5.0mm	
Output voltages	Open circuit:	
	51-46v	70-74v
Output	**50 volts**	**80 volts**
Welding current	Maximum hand welding	
	250 amps	195 amps
	Minimum hand welding	
	55 amps	35 amps
Maximum continuous hand welding current	110 amps	65 amps
Input voltage	240/415v	50 Hz 1 ph
Input	**240 volts**	**415 volts**
Input current	Maximum hand welding	
	72 amps	42 amps
	Max. cont. hand welding	
	30 amps	15 amps
Dimensions	Length	
	470mm	18½ inches
	Width	
	362mm	14¼ inches
	Height	
	420mm	16½ inches
Weight	53 kgs	117 lbs

Trojan 250 air cooled arc welder

GRINDERS

The illustration shows the three basic types of grinder, which are available in a range of sizes, wheel diameters and motor powers.

A typical specification for each type is given below.

Operating information
Hand-held angle grinder

Wheel size	180mm diameter × 10mm thick
Spindle speed	8500 rpm
Motor power	1500 W, 1 phase
Weight	5.2 kg

Double-end bench grinder

Wheel size	200mm × 20mm
Wheel speed	355 rpm

Motor speed	3000 rpm
Motor power	3 Phase 2000 W
	1 Phase 1500 W

Double-end pedestal grinder

Wheel size	300mm × 35mm
Wheel centres	406mm
Wheel speed	27.5 m/s
Spindle height	900mm
Spindle speed	1705 rpm
Motor 3 phase	1.5 kW

Source
Buck & Hickman Ltd   Telephone 0742 766660
Bank House
100 Queen Street
Sheffield
S1 2DW
UK

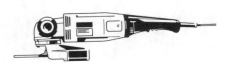

PNEUMATIC GRINDER

This simple machine has a high power-to-weight ratio and is suitable for redressing drill steels and bits. A line flow governor is used to regulate the torque supplied by the pneumatic motor to the grinding wheel.

Operating information

Type	40TC
Power	Compressed air supply required
Speed	3000 rpm unloaded at 5.6 Kg/cm^2
Wheel diameter	203mm

Air inlet	½ inch BSP connector
Length	550mm
Width	380mm
Height	285mm
Weight	17.6 Kg

Source
Enquiries to registered and sales office:
Holman-Climax   Telephone 44-8648
 Manufacturing Ltd   44-9156
Dolphin Court
2nd Floor
7A Middleton Street
Calcutta-700 071
India

DRILLS

The illustration shows the three basic types of drill, which are available in a range of sizes, styles and motor powers.

A typical specification for each type is given below.

Hand-held drill
This machine is a variable speed percussion model.

Chuck mm	Full load rpm		Blows per minute		Watts	Weight kg	Voltage AC
	High	Low	High	Low			
10	1750	975	35000	19500	420	1.6	220/240

Bench drill
Capacity	13mm
Distance centre of chuck to column	150mm
Spindle travel	95mm
Table	215 × 215mm
Spindle speeds (4)	745, 1020, 1990, 4260 rpm
Motor	½ hp

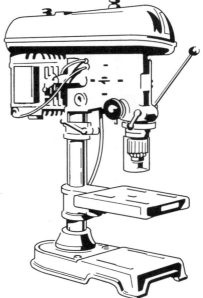

Pedestal drill
Drilling capacity in mild steel	40mm
Worktable size	400 × 400mm
Spindle traverse	250mm
Spindle speeds (6)	56-1000 rpm
Spindle feeds (4)	0.1-0.4mm per rev 0.004-0.016 in per rev
Spindle taper	4 MT
Motor	3 hp

Source
Buck & Hickman Ltd   Telephone 0742 766660
Bank House
100 Queen Street
Sheffield
S1 2DW
UK

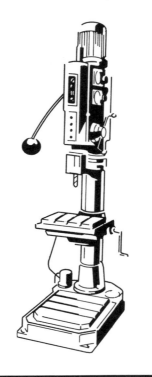

LATHE

The lathe illustrated is a heavy-duty gap-bed type which is available in a range of centre heights and distances between centres. The details below are for the smallest machine in this model range from the Gate Machinery Co. Ltd.

Operating information

Height of centres	12⅝″ (320mm)
Swing over bed	26⅜″ (670mm)
Swing over carriage	15¾″ (400mm)
Swing in gap	32½″ (825mm)
Width of gap	12¼″ (311mm)
Spindle bore	3¼″ (83mm)
Spindle speeds (18)	13 to 1000 rpm
Main motor	15 hp
Weights 1 metre model	3¼ tonne

Source

Gate Machinery Co. Ltd
BEC House
Victoria Road
London
NW10 6NY
UK

Telephone 081-965 0505
Telex: 264178

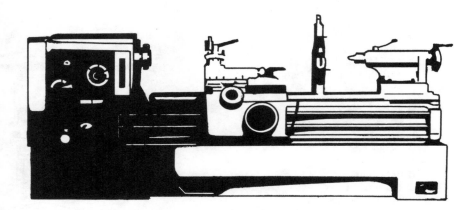

PRESS

In the lower capacity range it is possible to find manually-operated screw-type presses. However, for the type of maintenance work in the mining industry, a medium capacity hydraulic press is more appropriate. The 60-tonne machine illustrated would be adequate for most small maintenance workshops.

Operating information

Model PRS 60 tonne capacity floor press, hand operated:

Overall height	2035mm
Overall width	1560mm
Front clearance between posts	1060mm
Side clearance between posts	250mm
Ram diameter	80mm
Ram travel	220mm

Source

Buck & Hickman Ltd
Bank House
100 Queen Street
Sheffield
S1 2DW
UK

Telephone 0742 766660

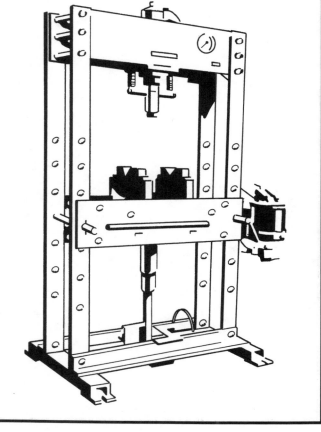

BENDING ROLLS

For any mine and treatment plant with a significant
potential platework maintenance problem, a set of
bending rolls is essential. The capacity and type
required depends upon the thickness of plate to be
handled and the volume of work. The range is from
light hand-operated machines through to large capacity
motor-operated units.

The machine shown in the illustration is a hand-
operated model for thin sheet steel.

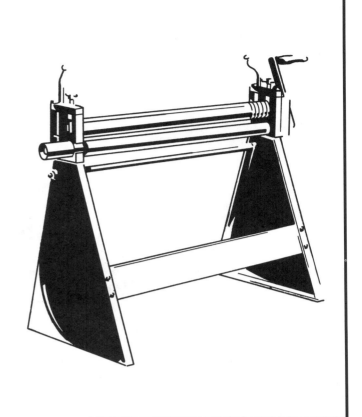

Operating information
Length of rollers   1010mm
Diameter of rollers   50mm
Thickness capacity   1.3mm

Source
Buck & Hickman Ltd   Telephone 0742 766660
Bank House
100 Queen Street
Sheffield
S1 2DW
UK

TROLLEY JACK

Source
Buck & Hickman Ltd   Telephone 0742 766660
Bank House
100 Queen Street
Sheffield
S1 2DW
UK

If any small vehicles (e.g. cars, trucks) are operated by
the mine, a trolley jack is an essential tool for the
vehicle mechanic. The normal capacity range is from
1 tonne through to 20 tonnes, and in general they are
similar to the illustration.

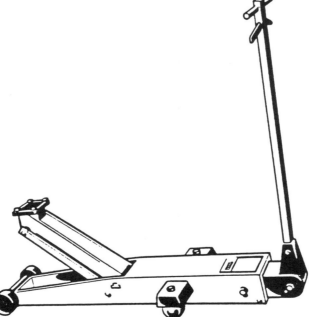

ELECTRIC HACKSAW

Designed for rapid heavy-duty cutting, these machines usually come equipped with a vice and a saw frame loading mechanism.

Specification

Capacity round and square	203mm
Capacity at 45° (with swivel vice)	143mm
Strokes per minute (2)	60 and 100
Motor	1.5 kW 2hp
Blade size minimum	350 × 32 × 1.6mm
Blade size maximum	425 × 32 × 1.6mm

Source

Buck & Hickman Ltd   Telephone 0742 766660
Bank House
100 Queen Street
Sheffield
S1 2DW
UK

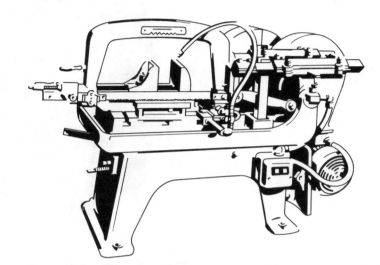

LARGE ELECTRIC MOTORS

We have received outline information from a manufacturer of electric motors, which are suitable for driving heavy mine plant and winders. Their product range starts large, and extends to sizes rated in megawatts that are unlikely to interest small-scale miners.

We are not able to illustrate an example here, but this company's motors are used in winders which they manufacture at Birmingham (UK) and one of these winders is illustrated in Section 5.

Operating information

There are three ranges of motors available:
Cage and wound-rotor induction motors, in sizes from 200 kW up to 25,000 kW.
Flameproof cage motors, from 100 kW up to 2,000 kW.
Motors for use in hazardous areas, again ranging in size right up to 25,000 kW.

Source

NEI Peebles Ltd   Telephone 031 552 6261
Peebles Electrical   Telex   72125
  Machines
East Pilton
Edinburgh
Scotland
EH5 2XT
UK

DIESEL ENGINE

The air-cooled engine illustrated here is one of a wide variety of engines which this company produces, ranging in size from 1.5 to 160 brake horsepower. The engine is fitted with an intake air cleaner and exhaust silencer.

Operating information

Cylinders	One
Cylinder capacity	408cc
Gross continuous power	4 brake horsepower
Drive shaft speed	1500 rpm
Net weight	83 Kg

Source

Lister/Petter Ltd   Telephone 0453 544141
Dursley   Telex   43261
Gloucester
GL11 4HS
UK

PORTABLE MOTOR GENERATOR

A generator such as this is ideal for supplying power to most resistivity and induced polarization geophysical surveys. It is supplied with a pack frame, enabling it to be carried by one man.

A four stroke Briggs and Stratton engine drives the generator, developing 5 HP at 3600 rpm. The output voltage is regulated by feedback from the survey transmitter.

Operating information

Output power	2 KVA
Output voltage	60V (45 to 80V)
Frequency	400 Hz (350 to 600 Hz), 3 phase
Dimensions	40 × 45 × 60cm, including frame
Weight	34 Kg

Source

Phoenix Geophysics Ltd	Telephone 416 477 8588
7100 Warden Avenue	Telex 06-986856
Unit 7	
Markham	
Ontario	
L3R 5M7	
Canada	

PORTABLE COMPRESSOR AND ELECTRICAL GENERATOR

An 11 HP petrol engine powers this unit, which produces 4000 Watts of electrical power and 8.9 ft³ per minute of compressed air at 100 psi.

A voltmeter and pressure gauge are fitted, and other fittings include a totally enclosed belt guard and a safety relief valve.

The unit is mounted on a hand pulled trolley fitted with two pneumatic tyred wheels. It is only one example of the units available from this manufacturer.

Source

Sepor Inc.	Telephone 213 830 6601
PO Box 578	Telex 194496
Wilmington	
CA 90748	
USA	

Operating information

Compressor	8.9 ft³ per minute at 100 lbs per square inch (0.25m³ at 7kg/cm²)
Generator	Three output sockets: — Two 120 Volts at 33.3 Amps — One 240 Volts at 16.5 Amps 50 or 60 Hz 4000 Watts maximum permissible load
Engine	11 HP Briggs and Stratton four stroke, hand started (electric start available as optional extra)
Length	48 inches (1.2m)
Height	33 inches (838mm)
Width	26 inches (660mm)
Weight	360 lbs (163 Kg)

ELECTRIC PORTABLE AIR COMPRESSOR

One of a wide range of compressors and associated air operated tools and equipment manufactured by this company, the unit described is powered by an electric motor which drives a water-cooled two cylinder compressor via a flexible coupling. The slow speed of rotation helps to keep the output air temperature low, and the compressor can be used in high ambient temperature conditions.

Two versions are available. One is configured in the conventional manner for surface working, mounted on a sprung chassis with pneumatic-tyred wheels, fitted with a folding drawbar. The alternative version has flanged wheels for rail track operation, and is flameproof for underground use.

Operating information

Type	T36EP
Power	66 kW (90 HP) 1000 rpm electric motor

Capacity	8.4m³ of free air per minute at 7 Kg/cm²
Air outlets	Four connectors, each ¾ inch BSP
Length	4.88m with drawbar folded
Width	1.83m
Height	1.93m
Weight	3000 Kg approximately

Source

Enquiries to registered and sales office:
Holman-Climax   Telephone 44-8648
  Manufacturing Ltd   44-9156
Dolphin Court
2nd Floor
7A Middleton Street
Calcutta-700 071
India

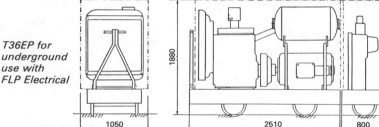

T36EP for underground use with FLP Electrical

1880   1050   2510   800

PORTABLE AIR COMPRESSOR

The compressor illustrated here is one of a wide range manufactured by this company in many sizes and supplied to all parts of the world. The principal use is for powering pneumatic rock drills and breakers.

This example is lightweight, and its low centre of gravity makes it stable when being towed. A lifting eye is fitted which allows placing by crane or helicopter.

Automatic overload protection shuts the compressor down if overheating occurs, sensed at the air outlet. The unit has a self-starter motor, and is fitted with an instrument panel which centralizes all necessary operational push buttons, monitoring instruments and indicator lamps.

The design is aimed at the minimum number of moving parts for longer life and low maintenance costs.

Operating information

Type illustrated	XAS 40
Working pressure	7 bar
Capacity	2.58m³ per minute free air delivery
Speed	1600 to 2000 rpm depending on load
Outlets	Two ball valve outlets are provided with quick coupling connections for attaching hoses
Power	Deutz 2 Hp two-cylinder air cooled engine
Length	2.7m
Width	1.53m
Height	1.295m
Dry weight	730 Kg

Source

Atlas Copco   Telephone 0442 61201
  (Great Britain) Ltd   Telex   825963
PO Box 79
Swallowdale Lane
Hemel Hempstead
Herts
HP2 7HA
UK

STATIONARY AND PORTABLE AIR COMPRESSORS

From this manufacturer a range of rotary vane compressors is available for internal combustion engine or electric operation. They are claimed to be exceptionally durable, of modern design, simple to install, maintain and adjust and supplying industrial quality compressed air with only 1 to 3 particles per million by weight. These Eisaire compressors come in two ranges, one for static mounting and the other on a two-wheeled towable chassis.

Operating information

Transportable models

Type	ROT 30	ROT 40	ROT 55	ROT 73	ROT 105	ROT 170
Capacity m³/min	3	4	5.5	7.3	10.5	17
Pressure Kg/cm²	8	8	8	8	8	8
Power in HP	34	46	65	82	112	173
Weight, KG	780	860	1030	1100	1620	2100
Height, mm	1250	1250	1500	1500	1500	1700
Width, mm	950	950	1150	1150	1150	1300
Length, mm	2500	2700	3200	3400	3900	4100

Static models

Type	EIR 25	EIR 40	EIR 50	EIR 60	EIR 100	EIR 150
Capacity m³/min	2.4	4	5	6	10	15
Pressure Kg/cm²	8	8	8	8	8	8
Power in HP	25	40	50	60	100	150
Weight, KG	530	670	850	960	1345	1550
Height, mm	930	930	1150	1150	1150	1150
Width, mm	950	950	1200	1200	1200	1200
Length, mm	1600	1700	2100	2200	2400	2600

Source

Equipos Industriales S.A. Telephone 921059
Av. Padre Claret 969
5147 Los Boulevares
Cordoba
Argentina

Eisaire static compressor

Eisaire transportable compressor

SUBMERSIBLE DRAINAGE PUMP

A portable and lightweight submersible pump for dewatering tasks on the surface or underground. Such pumps are in use worldwide in mining and construction.

The pump has a corrugated, galvanized-steel jacket for strength. Water cooling is provided for the motor when pumping but, when running dry, the motor is air cooled.

The electric motor has automatic overload protection, which will trip if either the temperature or current rises unduly (because of a blockage, for example).

The use of small pumps can avoid the need to install large central pumping stations because pumping can be carried out in stages.

Operating information
Motor supply	220V, 6.1A, 3 phase
Power	1.1 kW
Motor speed	2800 rpm
Weight	17 Kg
Height	480mm
Diameter	210mm

Source
Grindex   Telephone 08-74534
Box 538   Telex 17286
Hantuerkavagen
S-13625 Handen
Sweden

VERTICAL SAND PUMP

This is a centrifugal pump which is designed to handle coarse abrasive or frothy slurries. This version has a built in sump, but other pumps are available for use with existing sumps.

The body is made from ductile iron, and all parts subject to wear are hard castings.

Bearings are kept above the slurry level, and are sealed. Should the pump run dry, no damage will result. No packing glands or sealing water are needed, and the pump has been designed to avoid air locks.

The gravity feed to the runner enables the pump to handle varying feed rates.

Operating information
Motor	TEFC 1.5 or 2.0 HP
Capacity	
1.5 HP	0 to 30 qpm with 25 per cent solids at 30 feet TDH
2.0 HP	0 to 30 qpm with 25 per cent solids at 40 feet TDH
Drive	Vee belt

Source
Sepor Inc.   Telephone 213 830 6601
PO Box 578   Telex 194496
Wilmington
CA 90748
USA

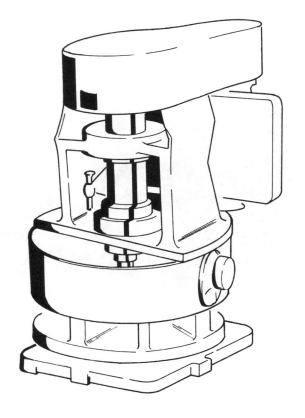

PORTABLE SELF-PRIMING CENTRIFUGAL PUMP SET

This is a small centrifugal pump which can be supplied with several drive options. In addition to electric motors, either a petrol or diesel engine can be specified, which makes the pump useful wherever simple water pumping tasks are needed. The company can also supply flameproof electric units for underground working, if required.

Castings are aluminium for the body and gunmetal for the impeller. Diesel and electric units are on a rigid bedplate for permanent mounting. A carrying handle is fitted to the petrol version, which is mounted on light channel-section base rails.

After initial priming the unit becomes self priming up to a head of about 7.5 metres for the engine-driven versions and 5.5 metres for the electric units. A check valve retains sufficient water in the pump body for restarting, but pumps may need repriming after long idle periods.

This supplier also makes a wide range of small to medium size centrifugal pumps for pressure boosting and other applications, and can supply associated hoses, strainers, priming tanks, safety valves, electrical starters and float switches.

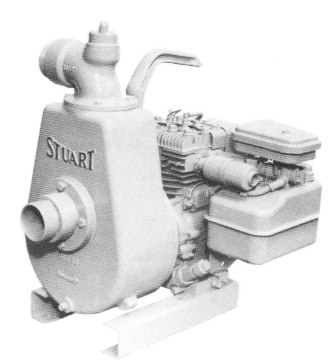

Petrol engine driven pump

Operating information

Pump type	Stuart SP2
Power options:	
Petrol	Briggs and Stratton 3 HP
Diesel	Petter 3.5 HP
Electric	2 HP 240 volt single phase 13 Amps or 400 volt three phase at 3.8 Amps. A starter with overload trip is required for the electric options.
Connections	2 inch BSP (approximately 50mm). The suction hose must be at least 50mm diameter, rise up to the pump and be kept as short as possible. A strainer at the intake end must be at least 50 per cent larger diameter than the pipe section to stop debris entering the pump. The 50mm outlet hose should run as straight as possible to the discharge point.
Weight	Range from 23.5 Kg for the petrol version up to 58 Kg for the diesel option
Performance	Refer to the graph

Source

Stuart Turner Ltd
Henley-on-Thames
Oxford
RG9 2AD
UK

Telephone 0491 572655
Telex 847093

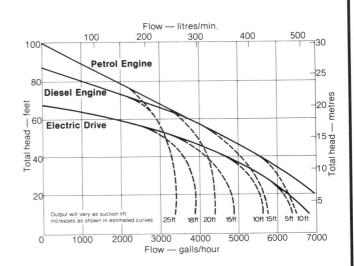

ABRASION AND CORROSION RESISTANT PUMPS

This manufacturer produces a wide range of pumps for many applications as well as equipment for mineral processing.

Operating information
The figures given relate only to the largest or highest performance model in each category. For details of smaller pumps please consult the manufacturer.

Slurry pumps
Flow rates up to	3500 litres per second
Heads up to	95 metres per stage
Pressures up to	6900 kPa (1000 lbs/in^2)

Vertical sump pumps
Flow rates up to	400 litres per second
Heads up to	40 metres

PC pumps
Construction	All heavier than conventional process pumps
Flow rates up to	400 litres per second
Heads up to	130 metres

Non-clogging cyklo pumps
Flow rates up to	120 litres per second
Heads up to	45 metres

Multistage turbine pumps
Flow rates	From 5 to 700 litres per second
Heads up to	300 metres

Source
Warman International Ltd   Telephone 286 0378
                                                286 8819
South East Commercial Co. Ltd
269/2-3 Liab Mae Nam Road
Chongnonsee
Bangkok 10120
Thailand

The above office supplied data, but Warman International have offices, agents or licensees in most countries. For example:

Warman International Ltd   Telephone 0706 814251
                                      Telex   63324
Halifax Road
Todmorden
Lancs
OL14 5RT
UK

Outer Casing
Split outer casing halves of ductile iron contain the wear liners. Casing halves provide full structural strength thus allowing inner liners to be fully worn before replacement becomes necessary.

Impeller
The impeller may be either moulded elastomer or metal. Deep side sealing vanes relieve seal pressure and minimize recirculation

Interchangeable Hard Metal and Moulded Elastomer Liners
Replaceable liners are available in both metals and pressure moulded elastomers and are fully interchangeable within the same pump.

Slurry pump construction

WATER PUMPS

These pumps are designed for general water circulation and pressure boosting duties. Using only three basic modules the company produces 22 different sizes, to cope with a wide range of flow and head conditions.

The pump casing is in two parts, bolted together. This allows easy dismantling for repairs and component replacement.

This company specializes in water pumps, and in pumps for handling non-abrasive slurries.

Operating information

Drive	Supplied as bareshaft units or as pumpsets with electric motor (as illustrated) in 50Hz or 60Hz versions
Pressure	16 bar standard working pressure
	20 bar maximum special working pressure
	14 bar maximum suction, according to pump model
Temperature	Up to 80 degrees Celsius with standard packing or seals
	Up to 120 degrees with optional packing or seals
	Down to minus 8 degrees Celsius

Source

SPP Pumps Ltd
Theale Cross
Reading
RG3 7SP
Berkshire
UK

Telephone 0734 323123
Telex   848189

This company has published a helpful pocket booklet for pump users. Principally intended for environmental systems, fire systems, industrial systems and oil and energy systems, this little booklet contains generally useful data and guidance in connection with the design of pumped water systems. The booklet is titled *Data for Pump Users*, and the 1988 edition is available post free from the above address at £4.50.

ABRASION AND CORROSION RESISTANT PUMPS FOR AFRICAN MINES

The Warman sump pump illustrated is similar to those listed in the last but one entry, but this manufacturer (in spite of the product name) is a different company from Warman International and exports principally to Africa (although they will handle enquiries constructively from readers in other countries).

Other Warman pumps for export to Africa by this company include many sizes and types of slurry pumps, dredge pumps, gravel pumps, froth pumps, solution pumps, electro-submersible pumps and submerged gland pumps.

Operating information

The following is Manwar International's selection chart for Warman sump pumps type SPR and SP.

Source

Manwar International   Telephone 078571 4960
 Ltd   Telex   367309
Lyme Hill Industrial Estate
Penkridge
Staffordshire
ST19 5LS
UK

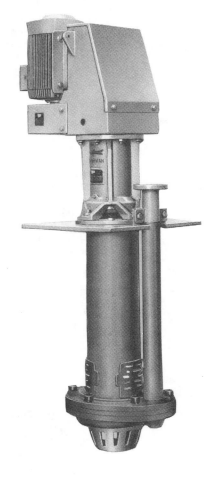

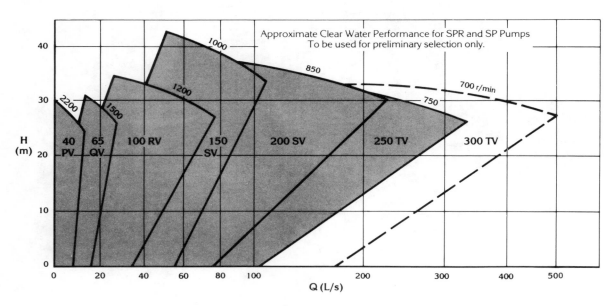

Approximate Clear Water Performance for SPR and SP Pumps
To be used for preliminary selection only.

Index

Index of suppliers